Single pour industrial floor slabs

Specification, design, construction and behaviour

John Knapton

Thomas Telford

Published by Thomas Telford Publishing, Thomas Telford Limited,
1 Heron Quay, London E14 4JD.

URL: http://www.t-telford.co.uk

Distributors for Thomas Telford books are
USA: ASCE Press, 1801 Alexander Bell Drive, Reston, VA 20191-4400
Japan: Maruzen Co. Ltd, Book Department, 3–10 Nihonbashi 2-chome, Chuo-ku, Tokyo 103
Australia: DA Books and Journals, 648 Whitehorse Road, Mitcham 3132, Victoria

First published 1999

A catalogue record for this book is available from the British Library

ISBN: 0 7277 2734 6

Typeset by MHL Typesetting Ltd, Coventry.
Printed and bound in Great Britain by Redwood Books, Trowbridge, Wiltshire.

Preface

This publication is a revision of a Guide originally prepared by the Author at the request of the NUFLOOR (Newcastle University Industrial Floor Research) Committee. NUFLOOR was established in 1993 by parties involved in the industrial floor construction industry with an interest in floors constructed using laser controlled screeding machines. It was felt that previous guidance was directed towards traditional floor construction methods. In recent years, the growth in the use of laser controlled screeding machines has led to a large number of single pour formless floors in which joints are introduced by cutting slots in the hardening concrete. Laser controlled screeding often leads designers to specify concrete including steel or polypropylene fibres. Such floors may behave in a manner which is different from that of traditional floors and these differences have been taken into account in this publication. The following organizations participated in the original work of NUFLOOR:

- Armorex Limited
- Bekaert Building Products
- English Partnership
- Fibermesh Europe
- RT James Limited
- Mowlem Northern
- Newcastle University
- Redland
- Snowden Flooring
- Laser Screed Limited

Details of Author

John Knapton is Professor of Structural Engineering at the University of Newcastle upon Tyne where he pursues his research interests into industrial floors, heavy duty paving, pavements surfaced with pavers, highway pavements, port pavements and aircraft pavements. Before taking the Newcastle Structures Chair, he ran his own consulting practice and was involved in the design and construction of many industrial floors.

His early career included spells as the British Constructional Steelwork Association Research Associate at Newcastle, Research Engineer in Cement & Concrete Association's Construction Research Department at Wexham Springs and Lecturer in Structural Engineering at Newcastle. He has represented the Institution of Civil Engineers on BSI committees and has published over 100 papers since 1974. His work in developing the Ghanaian rural village of Ekumfi-Atakwa has led to his enstoolment as Nana Odapagyan Ekumfi I (Chief Eagle of Ekumfi I) by the Ghanaian government's House of Chiefs.

Contents

Introduction

Automatically laser controlled screeding machines have been introduced into the UK in order to increase the quality of industrial floors and to speed up their construction. Their introduction has encouraged the use of fibre reinforced concrete since the placing of conventional mesh immediately in front of the laser screeder militates against the high levels of production achievable with laser screeders. Steel fibres not only govern joint spacings in the same way as mesh but additionally increase the strength of the concrete so permitting the use of thinner floors. Polypropylene fibres increase concrete toughness and durability but do not influence joint spacings or slab thickness. For these reasons, automatically laser controlled screeding of concrete floors has an impact upon design as well as construction. This book explains how laser screeders have changed floor construction and design. It provides specification and design guidance to allow a floor to be constructed to meet loading, durability and surfacing requirements.

In traditional long strip or chequerboard ground floor construction, slab thickness is governed by loads, concrete strength and ground conditions whilst the joint arrangement is governed by the type and quantity of reinforcing mesh. With fibre reinforced concrete, both slab thickness and joint configuration depend upon concrete properties. Also, in traditional construction, there is a construction related limit on maximum joint spacing. In laser screeded floors, joint spacing becomes a design variable and our preconceptions have been challenged. In this book, joint spacings can be selected according to structural behaviour.

A set of ten Design Charts that allow laser screeded slabs to be proportioned is included. The charts introduce the concept of Single Equivalent Wheel Load, whereby combinations of point loads of differing values and contact areas can be converted into a single load value and then used to choose a slab thickness. Illustrative examples of design for a point load and a distributed load are shown.

Chapters 1 and 2 present an overview of flooring materials and construction, with emphasis on materials and methods normally associated with laser screeded floors. In particular, polypropylene and steel fibres are described since it is common for them to be included. Working methods, specification data and design strengths are provided for commercially available fibres. Chapter 3 reviews some of the categories of loading commonly occurring on floor slabs and deals in particular with the effects of high-density storage systems on floor slab design. Some of the Design Charts described in Chapter 4 relate specifically to fibre reinforced concrete and the floor slabs monitored in Chapter 5 incorporated fibres. Chapter 5 presents data showing how joints in slabs open over the initial life of the slabs. The data can be used as a basis for selecting joint sealing material and can be used as an indication as to how a floor slab can be expected to behave. The data confirm that the recommendations presented in this book lead to slabs in which joints form correctly and then operate in a manner to reduce stresses in concrete.

1. Materials

1.1 Concrete

1.1.1 Introduction

Concrete is a man made composite material comprising natural aggregate, water and cement to bind the aggregates together to form a hard composite material. For most applications, concrete is defined or specified by its 28 day characteristic crushing strength, cement content and free water/cement ratio. When determining the load bearing capacity of a floor slab, the flexural strength of the specified concrete is needed. It is therefore customary to relate flexural strength to the characteristic crushing (or cube) strength of concrete. The main structural element of a ground bearing floor is the slab which is constructed over the sub-base and subgrade material, and careful consideration in the specification and quality control of the concrete should be ensured.

1.1.2 Specification

To specify concrete it is necessary to select its characteristic strength together with any limits required on mix proportions, the requirements of fresh concrete and the type of materials that may be used. It is also important to have a good understanding of the methods of transport, placing and compaction procedures that are to be used when specifying a concrete mix as this can considerably change the characteristics and performance of a slab. It is current practice to specify a 'designed mix' or special 'prescribed mix' and, where applicable, compliance testing procedures should be performed. BS 5328: 1981, *Methods of specifying concrete*,[1.1] defines the two types of mixes as follows.

> *A designed mix is specified by its required performance in terms of strength grade, subject to any special requirements for materials, minimum or maximum cement content, maximum free water/cement ratio and any other properties required. Strength testing forms an essential part of compliance.*
>
> *A prescribed mix is specified by its constituent materials and the properties or quantities of those constituents to produce a concrete with the required*

performance. The assessment of the mix proportions forms an essential part of the compliance requirements. Strength testing is not used to assess compliance.

A prescribed mix should be specified only when there is reliable previous evidence or data established from trial mixes, that with the materials and workmanship available the concrete produced will have the required strength, durability and other characteristics. This type of mix may be required to produce concrete having particular properties, e.g. to obtain a special finish.

1.1.3 Types of cementitious material

Commonly specified cementitious materials are listed below with their appropriate British Standard references.

BS 12: 1978 *Ordinary and rapid hardening Portland cement*
BS 146: 1973: Part 2 *Portland-blast furnace cement*
BS 1370: 1979 *Low heat Portland cement*
BS 3892: 1982: Part 1 *Pulverised-fuel ash for use as a cementitious component in structural concrete*
BS 4027: 1980 *Sulphate-resisting Portland cement*
BS 6588: 1985 *Portland pulverised fuel ash cement*
BS 6699: 1986 *Ground granulated blast furnace slag (ggbs) for use with Portland cement*

Ordinary or rapid hardening Portland cement is the most common cementitious material used but other cements or other combinations of Portland cement with ground granulated blast furnace slag (GGBS) and pulverized fuel ash (PFA) may be used provided that satisfactory data on their suitability, such as previous performance tests, are available. It has been suggested[1.2] that the replacement of up to 35% cement by PFA, or 50% GGBS, could be undertaken without an adverse effect on the wear resistance or flexural strength of the concrete slab, provided thorough curing for at least seven days has been performed. In the case of GGBS the flexural strength exhibited by the concrete may be enhanced.

1.1.4 Aggregates

Aggregates should comply with BS 882: 1992 *Aggregates from natural sources for concrete.*[1.3] The following recommendations should be complied with if a wearing surface is to be achieved. The physical/mechanical properties defined are determined in accordance with BS 812 and stated in BS 882:

- 10% fines value should not be less than 100 kN.
- Aggregate impact value should be not less than 30%.
- Flakiness index should not exceed 35.
- Drying shrinkage of concrete should be less than 0.065%.

Fine aggregate. The fine aggregate is usually naturally occurring sand. Aggregate passing a 5 mm BS 410 test sieve is termed sand. BS 882: 1992: Section 5.2.2 gives the grading limits of sand used for the construction of concrete floors. Sands within grading C or M should be used. Sand may be specified as either uncrushed, crushed or blended. Uncrushed sand results from the natural disintegration of rock, whereas crushed sand is the product of crushing processes of gravel or rock. Blended sand is a controlled mixture of two or more of the types described above. Very coarse or very fine gradings as well as gap gradings should not be used as this can often lead to difficulties in finishing or poor durability of the surface. The sand should be free from soft materials, such as soft sandstone, limestone, coal and lignite and the use of unwashed crushed fines can seriously inhibit the quality of the slab.

Coarse aggregate. For most slabs 20 mm maximum size aggregate is suitable, but for the construction of thicker slabs it may be more economical to use aggregates up to 40 mm maximum size. Gradings as defined in BS 882[1.3] should be used. Soft sandstone or soft limestone should be avoided and crushed igneous or crushed flint gravels of angular shapes are preferred. Aggregates other than those stated in BS 812: Part 1 may be used provided satisfactory data on the exhibited properties of concrete made with them are available. Recent research[1.4] has found that the use of angular or crushed aggregate has resulted in an increase in flexural strength of up to 25% as compared to the value obtained with rounded or irregular aggregates.

1.1.5 Admixtures

It is common practice to specify admixtures to aid the workability of fresh concrete without loss of strength or durability. Admixtures are permitted in designed and prescribed mixes and, if specified, they should comply to BS 5075: Part 1[1.5] and Part 3[1.6] as appropriate. The use of an additive will normally be determined by the Contractor according to his or her method of construction. If a construction procedure such as laser screeding requires a self levelling concrete, no water should be added to the mix on site. In this case the use of a super-plasticizer is common. A super-plasticizer is defined in BS 5075: Part 3[1.6] as an *admixture that, when added to a hydraulic binder concrete, imparts very high workability or allows a large decrease in water content for a given workability.* If two or more admixtures are to be used simultaneously, care should be taken to assess their interaction and to ensure their compatibility.

1.1.6 Concrete quality

Unlike most structural applications of concrete, floor slab design is based on the flexural strength of the concrete. A relationship between the 28 day characteristic compressive strength and the flexural strength of concrete is required. The 28 day characteristic compressive strength (in N/mm^2) is defined as the grade of the concrete and is prefixed by the letter C. A relationship between the flexural strength and the 28 day compressive strength is given in Table 1.1.[1.7]

From Table 1.1 the relevant flexural strength can be determined and factors of safety can be applied, resulting in the design flexural strength used in Chapter 4.

In order to obtain concrete of a particular strength there are a number of mix design limits to comply with. Developments in the technology of cement manufacture in recent times have resulted in the achievement of higher strengths from mixes than those achieved previously. Table 1.2 shows mix design guidance[1.8] for various grades of concrete. To achieve a durable wearing surface, concrete with a minimum cement content of 325 kg/m^3 and water/cement ratio not exceeding 0.55 is commonly specified, although stricter limits may be applied if the floor is to be subject to heavy industrial use.

The workability of fresh concrete should be suitable for the conditions of handling and placing, so that after compaction and finishing the concrete surrounds all reinforcement and completely fills its formwork. The workability of the concrete is normally determined by the Contractor to suit his or her method of working and a slump of 50 mm is the usual

Table 1.1. Relationship between concrete grade, compressive strength and flexural strength

Concrete grade Characteristic compressive strength: N/mm^2	C30 30	C40 40	C50 50
Flexural strength: N/mm^2	3.8	4.5	5.1

Table 1.2. Relationship between free water/cement ratio, cement content and lowest grade of concrete

Maximum free water/cement ratio	0.65	0.60	0.55	0.50	0.45
Minimum cement content: kg/m^3	275	300	325	350	400
Grade of concrete: N/mm^2	C30	C35	C40	C45	C50

maximum. If a slump of greater than 50 mm is allowed there is a tendency for the aggregate particles to segregate. Where the consistency of the mix is such that the concrete is unable to hold all its water, some is gradually displaced and rises to the surface. Separation of water from a mix in this manner is known as bleeding and can lead to dusting and poor wear resistance properties of the hardened concrete. The slump of a mix should therefore be carefully monitored on site during pouring.

The lower the water/cement ratio the higher is the strength of the hardened concrete (see Table 1.2), but this can lead to a loss in workability of the fresh concrete. Although modern equipment is capable of handling less workable mixes the use of super-plasticizers (see Section 1.1.5) to increase workability is now common. Lowering the free water/cement ratio in a concrete mix also has the advantage of allowing the finishing process to be started earlier.

1.1.7 Micro-silica concrete

Micro-silica concrete is specified for concrete floor construction where a tough marble-like wear resistant surface is needed. Tests commissioned by a manufacturer of micro-silica concrete yielded the results shown in Table 1.3.

The use of micro-silica concrete in floor slab construction results in an increase in flexural strength and thinner floors may be possible so compensating for some of the additional cost of the micro-silica. The recommended design flexural strengths of micro-silica concrete are given in Table 4.1.

Table 1.3. Test results showing the increase in 28 day compressive and flexural strength obtained from samples of micro-silica concrete as opposed to conventional C40 concrete

	Conventional C40 concrete	Micro-silica concrete
Cement content: kg/m^3	330	300
Water/cement ratio	0.55	0.45
28 day compressive strength: N/mm^2	55	83
28 day flexural strength: N/mm^2	5.9	7.7

1.2 Subgrade and sub-base

1.2.1 Subgrade

Subgrade is the naturally occurring ground or imported fill at formation level. Homogeneity of the subgrade strength is particularly important and avoiding hard and soft spots is a priority in subgrade preparation. Any subgrade fill should be suitable material of such grading that it can be well compacted. Fill containing variable piece sizes often proves difficult to compact, giving rise to settlement and early failure of the floor. On very good quality subgrades, such as firm sandy gravel, the sub-base material may be omitted.

1.2.2 Sub-base

The sub-base is the foundation to the floor slab. For most types of subgrade, a sub-base is essential. This layer usually consists of an inert, well graded granular material (see Section 1.2.5) or cement-treated material such as lean concrete or cement-bound granular material. *In situ* cement stabilization may prove to be an economical means of improving a poor subgrade. In the case of wheel and rack loading (see Chapter 3), the sub-base assists in reducing the vertical stress transmitted to the subgrade. Where a distributed load is present the floor slab achieves very little load spreading and the bearing capacity of the underlying subgrade may limit the maximum load applied to the floor.

1.2.3 Modulus of subgrade reaction

In considering the value of the stresses induced in a slab under loading, the influence of the subgrade is treated as that of an elastic medium with a modulus of subgrade reaction (K). Modulus of subgrade reaction characterizes the deflexion of the ground and/or the foundation under the floor slab. California Bearing Ratio (CBR) tests and plate bearing tests can be used to establish values (see Section 1.2.4). In many instances subgrades are variable and results obtained from *in situ* tests can often show scatter. CBR and plate bearing tests induce a shallow stress bulb and may not reflect the influence of deeper material which might become stressed beneath a loaded slab. Assumed values of the modulus of subgrade reaction K are shown in Table 1.4. Because the stresses in a concrete floor slab are insensitive to changes in the strength of the supporting material, the values in Table 1.4 may be assumed if no plate bearing test or CBR test results are available.

Chandler and Neal[1.7] suggest that the sub-base can be taken into account by enhancing the effective modulus of subgrade reaction K as in Table 1.5.

Table 1.4. Assumed modulus of subgrade reaction (K) for typical British soils[1.7]

	Typical soil description	Subgrade classification	Assumed K: N/mm^3
Coarse grained soils	Gravels, sands, clayey or silty gravels/sands	Good	0.054
Fine grained soils	Gravely or sandy silts/clays, clays, silts	Poor, very poor	0.027, 0.013

Table 1.5. Enhanced value of K when a sub-base is used

K value of subgrade alone: N/mm^3	Enhanced value of *K* when used in conjunction with:							
	Granular sub-base of thickness (in mm)				Cement-bound sub-base of thickness (in mm)			
	150	200	250	300	100	150	200	250
0.014	0.018	0.022	0.027	0.033	0.045	0.063	0.081	0.106
0.027	0.034	0.038	0.044	0.051	0.075	0.104	0.137	—
0.054	0.059	0.065	0.072	0.081	0.125	0.175		
0.082	0.089	0.096	0.105	0.114				

When a lean concrete sub-base is specified, the value of *K* of the subgrade material is used to calculate the required thickness of the concrete slab. This calculated thickness is then apportioned between the structural slab thickness (the higher strength concrete) and the lean concrete sub-base thickness. This relationship is shown in Table 1.6 when a C40 concrete is used for the slab and a C20 lean concrete is used for the sub-base.

Table 1.6. The modified thickness of a concrete slab with a C20 lean concrete sub-base

Calculated thickness of slab: mm	Modified thickness of slab (mm) when used in conjunction with lean concrete sub-base of thickness (mm):		
	100	130	150
250	190	180	—
275	215	200	—
300	235	225	210

1.2.4 Plate bearing and CBR testing

The plate bearing test procedure is to load the ground through a steel disc, usually mounted on the back of a vehicle, and to record load and corresponding deflexion. The value of K is found by dividing the pressure exerted on the plate by the resulting vertical deflexion and is expressed in units of N/mm^3, MN/m^3 or kg/cm^3. K is established by plate bearing tests with a plate loading diameter of 750 mm. A modification is needed if a different plate diameter is used: for a 300 mm diameter plate K is divided by 2.3 and for a 160 mm diameter plate it is divided by 3.8. Alternatively the CBR can be measured and CBR values expressed as percentages are obtained. The CBR of a soil is determined by a penetration test which measures the force required to produce a given penetration in the material. This force is compared with the force required to produce the same penetration in a standard crushed limestone. The result is expressed in percentage terms as a ratio of the two penetration forces. Thus a material with a CBR value of 4% offers 4% of the resistance to penetration as compared to that offered by standard crushed limestone. The laboratory test should be carried out in accordance with BS 1377.[1.9] Different subgrade materials will have different CBR values, and a conservative value is used for each category of soil. It is unusual for the CBR to be measured directly since it can usually be determined with sufficient accuracy from Liquid Limit (LL) and Plasticity Index (PI) values. If the CBR is to be measured directly, it should be done at the most adverse moisture content which the soil can reasonably be predicted to sustain. BS 1377 includes a 72 h soaking procedure which will be appropriate in some design situations. Table 1.7[1.10] shows the relationship between CBR and modulus of subgrade reaction K for a number of common soil types.

Table 1.7. Modulus of subgrade reaction and CBR values for a number of common subgrade and sub-base materials

	CBR: %	Modulus of subgrade reaction K: N/mm^3
Humus soil or peat	<2	unacceptable
Recent embankment	2	0.01–0.02
Fine or slightly compacted sand	3	0.015–0.03
Well compacted sand	10–25	0.05–0.10
Very well compacted sand	25–50	0.10–0.15
Loam or clay (moist)	3–15	0.03–0.06
Loam or clay (dry)	30–40	0.08–0,10
Clay with sand	30–40	0.08–0.10
Crushed stone with sand	25–50	0.10–0.15
Coarse crushed stone	80–100	0.20–0.25
Well compacted crushed stone	80–100	0.20–0.30

The standard method of classifying soils in the US for engineering purposes is the Unified System and this system can also be used to assess CBR and *K*. The Unified System classifies soils on the basis of grain size and plasticity. The initial division of soils is based on the separation of coarse (sand) and fine (clay) grained soils and highly organic soils (peat). The distinction between coarse and fine grained soils is determined by the amount of material retained on a No 200 (75 micron) sieve. Coarse grained soils are subdivided into sands and gravels on the basis of the amount of material retained on a No 4 (6 mm) sieve. Gravels and sands are then classed according to whether fine material is present. Fine grained soils are subdivided into two groups on the basis of LL and PI values. Where the soil has been classified in this way, it may be more convenient to use the *K* values below rather than the ranges in Table 1.7. The classification system subdivides soil types into different groupings according to the following system.

GW Well graded gravels and gravel – sand mixtures, little or no fines, $K > 0.082\,\text{N/mm}^3$

GP Poorly graded gravels and gravel – sand mixtures, little or no fines, $K > 0.082\,\text{N/mm}^3$

GM Silty gravels, gravel – sand mixtures, $K = 0.082\,\text{N/mm}^3$

GC Clayey gravels, gravel – sand – silt mixtures, $K = 0.054\,\text{N/mm}^3$

SW Well graded sands and gravely sands, little or no fines, $K = 0.054\,\text{N/mm}^3$

SP Poorly graded sands and gravely sands, little or no fines, $K = 0.054\,\text{N/mm}^3$

SM Silty sands, sand – silt mixtures, $K = 0.054\,\text{N/mm}^3$

SC Clayey sands, sand – clay mixtures, $K = 0.054\,\text{N/mm}^3$

ML Inorganic silts, very fine sands, rock flour, silty or fine sands, $K = 0.027\,\text{N/mm}^3$

CL Inorganic clays of low to medium plasticity, gravely clays, silty clays, lean clays, $K = 0.027\,\text{N/mm}^3$

OL Organic silts and organic silty clays of low plasticity, $K = 0.027\,\text{N/mm}^3$

MH Inorganic silts, micaceous or diatomaceous fine sands or silts, plastic silts, $K = 0.027\,\text{N/mm}^3$

CH Inorganic clays of medium to high plasticity, $K = 0.014\,\text{N/mm}^3$

PT Peat, mud and other highly organic soils, unacceptable

In the above list *G* = *gravel, S* = *sand, C* = *clay, W* = *well, P* = *poor, M* = *medium, H* = *high plasticity, L* = *low plasticity, O* = *organic, PT* = *peat* and *K* values have been rationalized to four values and these values are used in the Design Charts. The Unified System allows soils to be classified from any geographic location into categories to which engineering properties can be assigned, for example particle size distribution, LL and PI. The various groupings of this classification system have been devised to correlate in a general way with the engineering behaviour of soils. This procedure provides a useful step in any field or laboratory investigation for geotechnical engineering purposes.

1.2.5 Granular sub-base material (Type 1 and Type 2)

The information provided in this section is based on the UK Department of Transport (DTp) publication *Specification for Highway Works, Series 800, Road Pavements – Unbound Materials*.[1.11] The material should comprise an approved durable granular material such as gravel, hard clinker, crushed rock or well burnt colliery shale, blended if necessary with sand or other fine screenings.

Blast furnace slag for use of sub-base materials should comply with BS 1047. Steel slag may be used provided it has been weathered and conforms to the requirements of BS 4987: Part 1. Materials other than slag when placed within 500 mm of cement bound materials or concrete products should have a water soluble sulphate content not exceeding 1.9 g of sulphate (expressed as weight of SO_3 per litre) when tested in accordance with BS 1377: Part 3.

1.2.5.1 DTp granular sub-base material Type 1. Unless evidence suggests that Type 2 materials will be suitable, all granular sub-bases should be constructed from Type 1 materials which can comprise crushed rock, crushed slag, crushed concrete or well burnt non-plastic shale. The material must lie within the grading envelope of Table 1.8 and must not be gap graded. The sub-base material is transported, laid and compacted without drying out or segregation. The material must have a 10% fines

Table 1.8. Grading requirements for granular materials

BS sieve size	Percentage by mass passing	
	Granular sub-base material Type 1	Granular sub-base material Type 2
75 mm	100	100
37.5 mm	85–100	85–100
10 mm	40–70	40–100
5 mm	25–45	25–85
600 micron	8–22	8–45
75 micron	0–10	0–10

value of 50 kN or more when tested to BS 812: Part 111 and an Aggregate Crushing Value (ACV) of less than 30 when tested to BS 812: Part 111. Additionally, the material should have a CBR of 80% or more.

1.2.5.2 DTp granular sub-base material Type 2. Type 2 granular materials are made up of natural sands, gravels, crushed rock, crushed slag, crushed concrete or well burnt non-plastic shale. The specification states that the material must lie within the grading envelope of Table 1.8 and not be gap graded. The material is transported, laid and compacted at a moisture content within the range 1% above and 2% below the optimum moisture content and without drying out or segregation. The material must have a 10% fines value of 50 kN or more when tested to BS 812: Part 111. Additionally, the material should have a CBR of 20% or more.

1.2.5.3 Compaction of granular materials. Unbound material up to 225 mm compacted thickness is spread and compacted in one layer so that after compaction the total thickness is as specified. The minimum compacted thickness should not be less than 110 mm. Where the layers of unbound material are of unequal thickness the lowest layer should be the thickest layer. Compaction of unbound materials is carried out by a method shown in Table 1.9. The surface of any one layer of material on completion of compaction and immediately before overlaying should be well closed, free from movement under compaction plant and from ridges, cracks, loose material, pot holes, ruts or other defects. All loose, segregated or otherwise defective areas should be removed to the full thickness of the layer, and new material laid and compacted.

1.2.6 Cement stabilized sub-bases

The information provided in this section is based upon the DTp manual *Specification for Highway Works, Series 1000, Road Pavements – Concrete and Cement Bound Materials.*[1.12]

1.2.6.1 Constituents. The cement in cement-bound materials must comply with the materials listed in Table 1.10 or the combinations in Table 1.11

The maximum proportions of GGBS with Portland cement should not be greater than 65% of the total cement content for cement-bound materials. The water content should be the minimum amount required to provide suitable workability to give full compaction and the required density.

Table 1.9. Compaction requirements for granular sub-base material Types 1 and 2

Type of compaction plant	Category[b]	Number of passes for layers not exceeding the following compacted thicknesses (in mm)[a]		
		110	150	225
Smooth-wheeled roller (or vibratory roller operating without vibration	Mass per metre width or roll:			
	over 2700 kg up to 5400 kg	16	Unsuitable	Unsuitable
	over 5400 kg	8	16	Unsuitable
Pneumatic-tyred roller[c]	Mass per wheel:			
	over 4000 kg up to 6000 kg	12	Unsuitable	Unsuitable
	over 6000 kg up to 8000 kg	12	Unsuitable	Unsuitable
	over 8000 kg up to 12 000 kg	10	16	Unsuitable
	over 12 000 kg	8	12	Unsuitable
Vibratory roller[d]	Mass per metre width of vibrating roll:			
	over 700 kg up to 1300 kg	16	Unsuitable	Unsuitable
	over 1300 kg up to 1800 kg	6	16	Unsuitable
	over 1800 kg up to 2300 kg	4	6	10
	over 2300 kg up to 2900 kg	3	5	9
	over 2900 kg up to 3600 kg	3	5	8
	over 3600 kg up to 4300 kg	2	4	7
	over 4300 kg up to 5000 kg	2	4	6
	over 5000 kg	2	3	5
Vibrating plate compactor[e]	Mass per square metre of base plate:			
	over 1400 kg/m^2–1800 kg/m^2	8	Unsuitable	Unsuitable
	over 1800 kg/m^2–2100 kg/m^2	5	8	Unsuitable
	over 2100 kg/m^2	3	6	10
Vibro-tamper[f]	Mass:			
	over 50 kg up to 65 kg	4	8	Unsuitable
	over 65 kg up to 75 kg	3	6	10
	over 75kg	2	4	8
Power rammer[g]	Mass:			
	100 kg up to 500 kg	5	8	Unsuitable
	over 500 kg	5	8	12

Table 1.10. Cementitious material specifications

Cement	Complying with
Portland cement (PC)	BS 12
Portland blast furnace cement (PBC)	BS 146
Portland pulverized fuel ash cement	BS 6588
Pozzolanic cement (Grades C20 or below)	BS 6610

Notes to Table 1.9

[a] The number of passes is the number of times that each point on the surface of the layer being compacted is traversed by the item of compaction plant in its operating mode (or struck in the case of power rammers).

[b] The compaction plant is categorized in terms of static mass. The mass per metre width of roll is the total mass on the roll divided by the total roll width. Where a smooth-wheeled roller has more than one axle, the category of the machine is determined on the basis of the axle giving the highest value of mass per metre width.

[c] For pneumatic-tyred rollers the mass per wheel is the total mass of the roller divided by the number of wheels. In assessing the number of passes of pneumatic-tyred rollers the effective width is the sum of the widths of the individual wheel tracks together with the sum of the spacings between the wheel tracks providing that each spacing does not exceed 230 mm. Where the spacings exceed 230 mm the effective width is taken as the sum of the widths of the individual wheel tracks only.

[d] Vibratory rollers are self propelled or towed smooth-wheeled rollers having means of applying mechanical vibration to one or more rolls. The requirements for vibratory rollers are based on the use of the lowest gear on a self propelled machine with mechanical transmission and a speed of 1.5–2.5 km/h for a towed machine. Vibratory rollers operating without vibration are classified as smooth-wheeled rollers.

[e] Vibrating-plate compactors are machines having a base plate to which is attached a source of vibration consisting of one or two eccentrically weighed shafts. They normally travel at speeds of less than 1 km/h

[f] Vibro-tampers are machines in which an engine-driven reciprocating mechanism acts on a spring system, through which oscillations are set up in a base plate.

[g] Power rammers are machines which are actuated by explosions in an internal combustion cylinder, each explosion being controlled manually by the operator. One pass of a power rammer is considered to have been made when the compacting shoe has made one strike on the area in question.

Table 1.11. Cementitious material combination specifications

Combination	Complying with
Portland cement with ground granulated blast furnace slag	BS 12
Portland cement with pulverized fuel ash for use as a cementious component	BS 3892: Part 1
Portland cement with micro-silica having a current BBA certificate	BS 12

Table 1.12. Grading of aggregate materials used in the four categories of cement bound materials

BS sieve size	Percentage by mass passing nominal maximum size:			
	–	–	40 mm	20 mm
	CBM1	CBM2	CBM3 & CBM4	
50 mm	100	100	100	–
37.5 mm	95	95–100	95–100	100
20 mm	45	45-100	45–80	95–100
10 mm	35	35-100	N/A	N/A
5 mm	25	25-100	25-50	35–55
2.36 mm	N/A	15–90	N/A	N/A
600 micron	8	8–65	8–30	10–35
300 micron	5	5–40	0–8	8–8
75 micron	0	0–10	0–5	0–5

1.2.6.2 Cement Bound Material Category 1 (CBM1). CBM1 is typically made from a material which has a grading finer than the limits in Table 1.12.

1.2.6.3 Cement Bound Material Category 2 (CBM2). CBM2 is typically made from gravel–sand, a washed or processed granular material, crushed rock, all-in aggregate, blast furnace slag or any combination of these. The constituents of the material must fall within the grading limits shown in Table 1.12. The material must have a 10% fines value of 50 kN or more when tested in accordance with BS 812: Part 111 with samples in a soaked condition.

1.2.6.4 Cement Bound Material Category 3 (CBM3). CBM3 is made from natural aggregate material complying with BS 882.

1.2.6.5 Cement Bound Material Category 4 (CBM4). CBM4 is made from natural aggregate material complying with BS 882. If blast furnace slag aggregate is to be used, it must comply with BS 1047: 1983. Cement for use in all cement-bound material and aggregate for use in CBM3 and CBM4 should be kept dry and used in the order in which it is delivered to the site. Different types of cementitious material must be stored separately.

1.2.6.6 Drylean concrete. Drylean concrete is a lean concrete with a low water content. The maximum aggregate to cement ratio is 15 to 1. The water content should be between 5% and 7% by weight of dry materials, the final value being selected to give the maximum dry density. The material should be rolled to give the maximum possible density.

1.2.6.7 Batching and mixing. Cement-bound materials should be made and constructed as summarized in Table 1.13. Batching and mixing

Table 1.13. Batching and mixing of cement-bound materials

Site requirements					Specimen requirements	
Category	Mixing plant	Methods of batching	Moisture content	Minimum compaction	Minimum 7 day cube compressive strength: N/mm^2	
					Average	Individual
CBM1	Mix-in-place or mix-in-plant	Volume or mass	To suit requirements for strength, surface level, regularity and finish	95% of cube density	4.5	2.5
CBM2	"	"	"	"	7.0	4.5
CBM3	Mix-in-plant	Mass	"	"	10.0	6.5
CMB4	"	"	"	"	15.0	10.0
Drylean Concrete	"	"	Between 5% & 7% of dry weight	Maximum possible	15.0 (Maximum) (No single cube below 12)	

should be carried out in the appropriate manner described in Table 1.13. Where the mix-in-plant method is used and materials are batched by mass, materials should be batched and mixed in compliance with BS 5328: Part 3.[1.13]

1.2.6.8 Transporting. Plant-mixed cement-bound material when mixed should be removed from the mixer immediately and transported directly to the point in consideration.

1.2.6.9 Laying. All cement-bound material should be placed and spread evenly in such a manner as to prevent segregation and drying. Spreading the material should be undertaken concurrently with placing or without delay. Base cement-bound material is often spread using a paving machine or a spreader box and operated with a mechanism which levels off the cement-bound material to an even depth. Cement-bound material is always spread in one layer so that after compaction the total thickness is as specified. Compaction is carried out immediately or within 2 h of the addition of the cement. The surface of any layer of cement-bound material on completion of compaction and immediately before overlaying should be well closed, free from movement under compaction plant and from ridges, cracks, loose material, pot holes, ruts or other defects.

1.2.6.10 Compaction. Compaction should be carried out immediately after the cement-bound material has been spread and in such a manner to prevent segregation. Compaction must be completed within 2 h of the addition of the cement. The surface of any one layer of cement-bound material on completion of compaction and before overlaying should be well closed, free from movement, compaction plant and from ridges, cracks, loose material, pot holes, ruts or other defects.

1.2.6.11 Curing. Immediately on completion of compaction, the surface of the cement-bound sub-base should be cured for a minimum period of seven days.

1.2.7 Settlement

A site investigation can provide the necessary information for an estimate of long-term settlement to be made. Where slabs are supported on subgrade such as organic soils, heavy clays and loose sands, or where land has been reclaimed, anticipated long-term settlements may be significant. Plate bearing tests, as described in Section 1.2.4, enable long-term settlements to be predicted. Soil stabilization, drainage or compaction, or piled foundations may be used to reduce or eliminate settlement.

1.3 Slip membranes

The slip membrane consists of polyethylene sheeting laid beneath the floor slab with overlaps of at least 200 mm. It is usually placed immediately prior to concrete pouring. Wrinkles and folds should be completely removed as they can result in weakening of the slab in later life as they may form crack inducers. It is advisable to anchor the polyethylene sheeting with small heaps of concrete, especially on the overlaps. A minimum of 125 micron (500 gauge) polyethelene sheeting should be used but 250 micron (1000 gauge) or 300 micron (1200 gauge) sheets are common.

A slip membrane is used to reduce friction between a concrete slab and its sub-base. The coefficient of friction with the use of a membrane is in the region of 0.2, compared with values of up to 0.7 when the concrete slab and sub-base are in direct contact. Prevention of loss of moisture and fines from the fresh concrete into the sub-base does occur, although a slip membrane is not intended or required to serve as a damp-proof membrane. When damp-proofing is to be provided thicker sheets or more elaborate measures may be required.

If an impermeable membrane is used then drying can take place only from the upper surface of the slab which may result in curling (see Chapter 5). The perforation of the slip membrane, or complete omission with the use of a blinding material, may therefore need to be considered although this would result in an increased loss of water and fines from the underside of the slab.

1.4 Fibres as reinforcing material

During recent years, floor construction methods involving the addition of polypropylene or steel fibres into a concrete mix have become common in the UK. For centuries, man has attempted to reinforce construction mortars and concretes with various types of fibres. Pharaoh ordered his foremen to cease supplying reinforcing straw to Israelite brick makers (Exodus 5[7]) and later the Romans used hair fibres in structural mortars.

With the advent of fast-track systems in the construction industry, concrete flooring has had to meet quicker construction programmes. With the use of laser screeders, fibres are often specified instead of conventional mesh because of the inconvenience of positioning individual mats of mesh immediately in front of the laser screeding machine as the machine progresses. Laser screeding machines cannot construct conventional long strip mesh reinforced floors efficiently as both the mesh and the formwork impede the machine. As a consequence, plain concrete or fibre based concrete are often specified for laser screeded floors.

1.5 Polypropylene fibres

1.5.1 Introduction

Polypropylene fibres for concrete can be in fibrillated or monofilament form manufactured in a continuous process by extrusion of polypropylene homopolymer resin. They are usually coated to improve wetting and dispersibility within the cement paste and to increase the extent of contact and bond between the fibres and the concrete matrix in the hardened state. Polypropylene fibres are not a substitute for conventional structural reinforcement or normal good curing procedures, but they may be used as an alternative to non-structural mesh (see Section 1.7) for crack control purposes acting as a secondary reinforcement. The design of polypropylene fibre based concrete floors proceeds as for unreinforced floors. The main purpose of polypropylene fibres is to provide crack control by distributing and absorbing tensile stresses which may occur as a result of shrinkage and temperature movements, particularly in the early life of the slab when the concrete has yet to reach sufficient tensile strength. They do not eliminate cracks and are not considered to contribute to the strength of the slab.

1.5.2 Monofilament fibres

Monofilament fibres are manufactured from extruded sheet/film material which is subject to molecular alignment, coated and cut to the appropriate length. This type of fibre is usually much finer than the fibrillated fibre (see Section 1.5.3) and the properties of a concrete resulting from the addition of monofilament fibres depend on the large number of fibres present. A smoother surface finish may be achieved from the use of monofilament fibre as opposed to the fibrillated type. Monofilaments do not provide any mechanical bond to the cement paste, but rely on their greater number per cubic metre of concrete and their chemical bond in order to achieve their proven qualities in both plastic and hardened states.

1.5.3 Fibrillated fibres

Fibrillated fibres are manufactured from extruded sheet/film material which is subject to molecular alignment, fibrillated, coated and cut to the appropriate length. Clustering of fibres is overcome by the mixing of aggregates in the concrete mix. Basic properties of fibrillated fibre[1.14] are:

- density = 900 kg/m^3
- tensile strength range = 560–770 N/mm^2
- elastic modulus = 3.5 kN/mm^2
- melt point = 160–170 °C.

Fibrillated fibres have a rough surface texture which gives each fibre a high degree of mechanical bond to the concrete. Monofilament fibres achieve enhanced plastic shrinkage control and trowel workability, while fibrillated fibres impart a higher degree of abrasion resistance to the resulting concrete.

1.5.4 Addition and mixing

The addition of polypropylene fibres is at a recommended dosage of 0.90 kg/m^3 (0.1% by volume or 1 l/m^3). They are compatible with all cementitious products and admixtures and generally require no change in mix design or water/cement ratio. The fibres may be added at either a conventional batching/mixing plant or by hand to the ready-mix truck on site. An even distribution throughout the concrete can be achieved in a 6 m^3 truck mixer in five minutes at full mixing speed.

1.5.5 Placing, curing and finishing

Concrete mixes containing polypropylene fibres can be transported by normal methods and flow easily from the hopper outlet. No special precautions are necessary when pouring and fibre-dosed concrete will flow around an obstruction such as reinforcement in the same manner as a conventional concrete mix of similar proportions. Conventional means of tamping or vibration to provide the necessary compaction can be used.

Curing procedures similar to those specified for conventional concrete should be strictly undertaken. If steam curing at a temperature in excess of 140 °C is to be used, polypropylene fibres should not be used. The fibres do not affect the hydration rate or stiffening time of the concrete.

Placed fibre-dosed mixes may be floated and trowelled using all normal hand or power tools. Occasional fibres protruding through the surface will quickly wear away. Workmanship should comply with the relevant requirements of BS 8204 Part 2: Sections 2.1 and 2.2.[1.15] Anti-wear products and other toppings may be used.

1.5.6 Controlling plastic shrinkage cracks in concrete

Plastic cracking may occur in the plastic concrete as a result of drying shrinkage. Plastic cracks are formed within the first 24 h after the concrete has been placed when the evaporation rate is high and the surface of the concrete dries out rapidly. Plastic shrinkage cracks generally pass through the entire slab and form weaknesses, permanently lowering the integrity of the floor slab before the concrete has had the opportunity to gain its design strength. Plastic cracks may occur through the whole depth of a slab and cannot be remedied by surface treatment.

Polypropylene fibres inhibit plastic cracking by holding water at or near the surface of the concrete, delaying evaporation and increasing cement hydration. Therefore, bleeding is inhibited. As concrete hardens and shrinks, micro-cracks develop. When the micro-cracks intersect a fibre strand, they are blocked and prevented from developing into macro-cracks and hence plastic cracking. This reduction of micro-cracks in the plastic state enables the concrete to better develop its optimum integrity. A number of research programmes studying the effect on plastic shrinkage cracking of concrete with the use of polypropylene fibres are described in Section 5.4.2.

1.5.7 Effect on workability

Polypropylene fibres act mechanically – the cohesive effect is largely due to surface tension and breaks down under vibration and compaction. The slump of a fibre-dosed concrete will be lower as a result of the thixotropic effect caused by the fibres but the mobility or placeability of the concrete is generally unaffected. Water should not be added to compensate for this thixotropic effect. Vebe and compaction factor tests are not significantly affected by the addition of polypropylene fibres. The improved cohesiveness also proves to be beneficial in pumped concrete owing to the reduction in rebound when placing.

1.5.8 Polypropylene fibre reinforced concrete

The use of polypropylene fibres in the construction of a ground floor concrete slab is not considered to contribute to the strength of the slab. The addition of polypropylene fibres at the usual recommended amount (0.9 kg/ m^3) will not significantly affect the ultimate compressive, tensile or flexural strength of the concrete matrix. Before ultimate stress is reached the performance of a fibre-enhanced concrete is improved in a number of ways. These improvements are due to concrete being an inherently variable material with a wide range of stress concentrations and the addition of the fibres favourably reduces this variability. If a fibre is aligned across a crack there is a small increase in stress required for crack propagation to occur.

1.5.9 Strength characteristics

Tests conducted by a manufacturer of polypropylene fibres[1.14] revealed the change in strength characteristics shown in Table 1.14.

Compressive strength tests conducted in accordance with BS 1881 indicated that the fibres, when used at the recommended dosage rate of 0.90 kg/m^3, slightly increase the early strength gain of concrete. The fibres have no significant effect on the 28 day compressive strength of

Table 1.14. Test results comparing the strength of polypropylene fibre-dosed concrete and conventional plain concrete

	Strength of fibre reinforced concrete: N/mm^2	Strength of unreinforced concrete: N/mm^2
Compressive strength (equivalent cube method)		
1 day	16.5	16.0
3 days	28.5	24.5
7 days	34.0	35.0
28 days	43.5	39.5
Cube compressive strength		
1 day	16.0	14.5
3 days	28.0	27.5
7 days	34.0	36.0
28 days	48.5	44.5
Flexural strength		
1 day	2.3	2.1
3 days	4.0	3.7
7 days	4.2	4.8
28 days	4.6	6.2

concrete cubes nor do they have any substantial effect on the flexural strength of the concrete.

1.5.10 Shatter resistance

Typically, when concrete test cylinders fail in compression at ultimate load, there is an initial crack. Continued loading with plain concrete specimens causes the cylinders to fragment and fall apart. With polypropylene fibres present the concrete specimen holds together after maximum load without falling apart or shattering. Tests[1.14] show the ability of polypropylene fibre-dosed concrete to remain intact and not to shatter after more than 10% additional compression as compared to plain concrete which shattered completely shortly after the first crack developed. This characteristic of polypropylene fibre-dosed concrete is important in applications where there are impact or seismic concerns.

1.5.11 Impact resistance

The addition of polypropylene fibres increases the energy absorption/impact of a concrete slab. The fibres bridge the cracks that develop, thereby inhibiting further crack growth. Therefore whereas the ultimate tensile strength of fibre-dosed concrete does not increase appreciably, the tensile strain at rupture does. Where steel reinforcement is used in concrete the addition of fibres enhances the bond between the concrete and the reinforcing bars by inhibiting cracking on the concrete under bearing stress.

1.5.12 Abrasion resistance

The introduction of polypropylene fibres into concrete results in a greater surface abrasion resistance compared to that of conventional concrete. Tests[1.14] have shown that the presence of fibres in a concrete mix reduces the amount of bleeding and assists in holding aggregate near the surface of fresh concrete, so resulting in better surface integrity. Appendix B shows the results of abrasion tests of fibrous concrete.

1.5.13 Permeability

Permeability is defined as the ease with which a fluid can flow through a solid. The addition of polypropylene fibres to a concrete slab reduces its water permeability owing to fibres interfering with the normal bleed channels and capillaries that are initially formed in the plastic state. With the reduction of cracking of the concrete resulting from the inclusion of fibres, the penetration of water has been laboratory tested[1.14] to be reduced by at least 50%. Figure 1.1 shows how increasing the polypropylene fibre dosage decreases the permeability of the concrete.

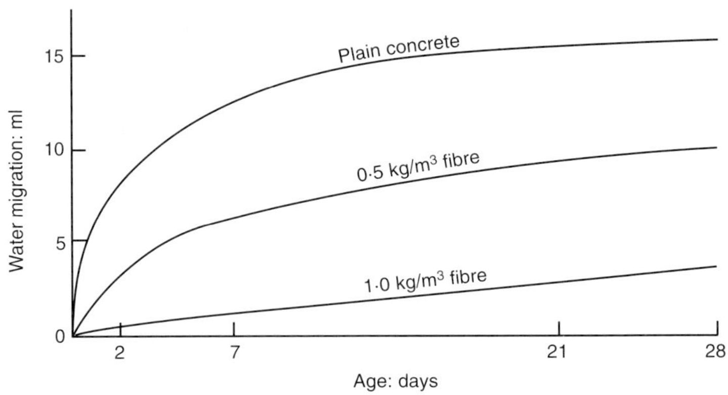

Fig. 1.1. The effect of polypropylene fibres on concrete permeability

1.5.14 Resistance to freeze/thaw

Fibre-dosed concrete has a significantly enhanced resistance to frost attack. There is some evidence to suggest that polypropylene fibres may be considered as an alternative to air-entrainment to obtain freeze/thaw resistance.

1.5.15 Chemical resistance

The presence of fibres does not alter the chemical resistance of concrete. Polypropylene is an inert and alkali resistant material and will not degrade in concrete.

1.5.16 Reduction of the corrosion of steel reinforcement

The addition of polypropylene fibres into a concrete slab significantly increases the protection of the steel reinforcement, within the slab, against corrosion. The reduction in permeability of the concrete is an attribute of prime importance with regard to the protection against corrosion. The high toughness index, which is the ability to sustain a load after initial crack, is also important in reducing the corrosion of reinforcing steel. This is due to the reduction of spalling of the concrete and the continued bond to the steel reinforcement.

1.5.17 Fire resistance

The inclusion of polypropylene fibres in a concrete mix does not affect the fire resistance of the finished concrete floor slab.

1.6 Steel wire fibres

1.6.1 Introduction

Steel fibres may be used in place of mesh reinforcement. The stresses that occur in a floor slab are complex and, depending on the type of load, tensile stresses can occur at the top and at the bottom of the slab. There are, in addition, stresses that are difficult to quantify, arising from a number of causes such as sharp turns from fork lift trucks, shrinkage and thermal effects and impact loads. The addition of steel wire fibres to a concrete slab results in a homogeneously reinforced slab achieving a considerable increase in flexural strength and enhanced resistance to shock and fatigue.

1.6.2 Concrete composition and quality

In order to obtain steel fibre reinforced concrete that is easy to pump and to work, with minimum shrinkage, a steel wire fibre manufacturer[1.10] specifies the following.

- Quantity of cement (commonly Ordinary Portland Cement) should be between 320 and 350 kg/m^3.
- 750 to 850 kg/m^3 good quality zero to 4 mm well graded sharp sand should be used.
- A continuous aggregate grading with a maximum size of 28 mm for rounded gravel and 32 mm for crushed stone should be used. Limit the fraction larger than 14 mm to 15–20%.
- Characteristic compressive strength of at least 25 N/mm^2 should be used.
- Water/cement ratio should be about 0.50, and should not exceed 0.55
- The use of a super-plasticizer is permitted to obtain the necessary workability.
- Admixtures of chloride or chloride-containing concrete additives are not permitted.

1.6.3 Addition and mixing

The recommended dosage rate of steel fibres is between 20 and 40 kg/m^3. The greater the dosage rate the greater is the flexural strength of the slab for a particular grade of concrete. Fibres can be added at the mixing plant or on site directly into the mixing truck. At the mixing plant steel fibres are usually added into the mixer at the same time as the aggregates. On site the concrete must first achieve the correct workability by the addition of a super-plasticizer before the fibres are added (see Section 1.1.5). The fibres should then be added at the manufacturer's specified rate resulting in a uniform distribution. For example, one manufacturer[1.10] recommends addition at a rate of two 30 kg bags per minute with the truck rotating at full mixing speed and mixing continuing for a further two minutes after the addition of the full dose. Visual inspection during pouring is necessary to check satisfactory fibre distribution. All fibre bundles must have separated into individual fibres for mixing to be sufficient.

1.6.4 Placing, curing and finishing

When placing a floor slab the concrete should be compacted as effectively as possible. Conventional means of tamping or vibration can be used. The usual techniques of floating and trowelling can be applied for finishing. After compaction and levelling, anti-wear products and cement are often spread on top of the concrete surface. Brushing of the concrete surface can be undertaken. The fresh concrete should be protected during curing periods by closing all openings to the building with the exception of ventilation holes. Immediately after finishing, a curing compound should be applied to combat rapid drying, forming an unbroken film on the surface of the concrete. A second curing layer may be applied if environmental conditions might cause rapid drying. Thin floors, of thickness 120 mm or less, should be provided with a double curing layer to prevent the risk of curling at the edges resulting from overfast drying. It should be noted, however, that if a wear resistant topping is to be included a curing compound must not be applied. In this case the concrete can be kept moist by wet spraying or by overlaying plastic sheeting. Plastic sheets, however, must not be applied if there is a risk that the temperature will become too high and result in the concrete setting too quickly.

1.6.5 Types of steel fibre available

The most commonly used steel fibre is the 60 mm long hooked fibre. Hooked fibres are usually glued together (collated) with a special water soluble glue to form fibre plates which readily disperse in the concrete mixer. The hooks help to ensure optimum fibre anchorage (or adhesion) in the hardened concrete. Enhanced adhesion can be achieved by either anchorage points at the ends of fibres (e.g. a pedal or hook) or in the case of a crimped fibre by adhesion along the whole length of the fibre. It is usual to consider only fibres with enhanced adhesion for reinforcement in concrete industrial floors. Various types of steel fibres are illustrated in Fig. 1.2.

Breaking or premature deformation of the fibres is prevented by the very high tensile strength of the drawn wire (usually greater than 1100 N/mm^2). The aspect ratio, which is the ratio of fibre length to fibre diameter, is also an important factor in fibre specification with common values of between 60 and 75.

1.6.6 Controlling cracking

Steel wire fibres effectively limit the extension of micro-cracks always present in concrete (see Fig. 1.3). In concrete without fibres, tension cannot be transmitted across the crack, that is, once the tensile capacity of the plain concrete is exceeded, the micro-crack will extend rapidly resulting in brittle failure. The action of steel wire fibres in a concrete slab is to reduce the concentration of stresses near the micro-cracks by:

(i) fibres bridging the crack and therefore transmitting some of the load across the crack;
(ii) fibres near the crack tip resisting more load owing to their higher modulus of elasticity compared to that of the surrounding concrete.

A crack is formed where the ultimate stress in the floor slab is exceeded locally. Steel fibres cause the crack to behave as a hinge, resulting in a redistribution of stresses. Unlike a broken zone in a brittle material, this

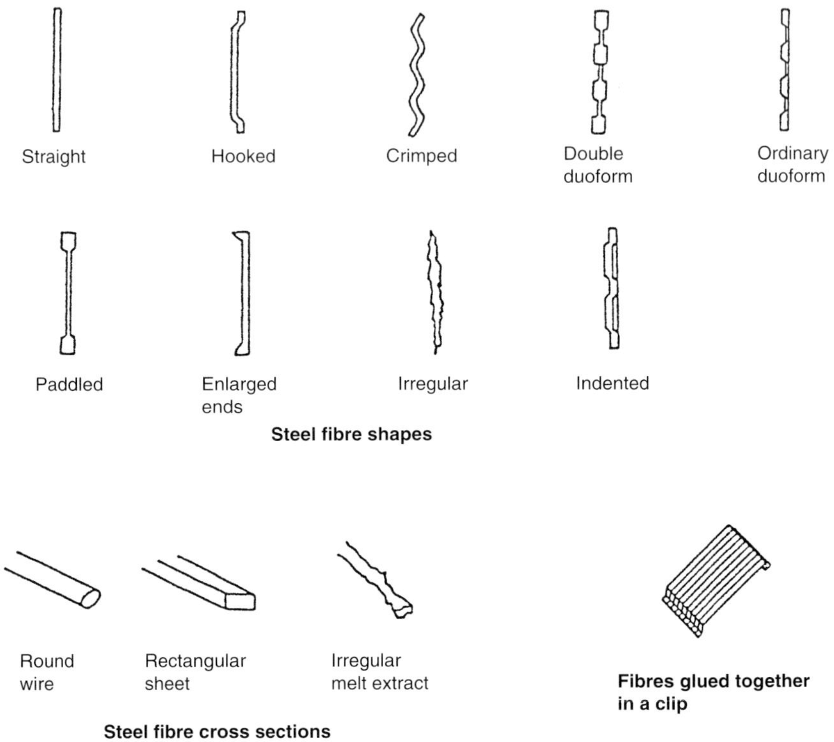

Fig. 1.2. Different steel fibre types

hinge can still resist stresses (depending on the type and dosage used) and thus the load bearing capacity of the floor is increased. Section 5.4.1 shows the results of a research programme on shrinkage cracking of steel fibre reinforced concrete. Steel wire fibres should not be specified to prevent micro-cracking. Micro-cracking in steel fibre reinforced concrete will occur at a similar rate to that expected in plain concrete, but the addition of steel fibres prevents micro-cracks from developing into macro-cracks.

1.6.7 Flexural strength properties

The primary function of introducing steel fibres into a concrete mix is to increase the load capacity of the slab. Unlike most structural applications of concrete, when designing a floor we rely upon the flexural strength or modulus of rupture of the concrete, to which we assign a design value. In the US flexural strength is defined as the stress corresponding to the occurrence of the first crack in the test specimen. This is the point at

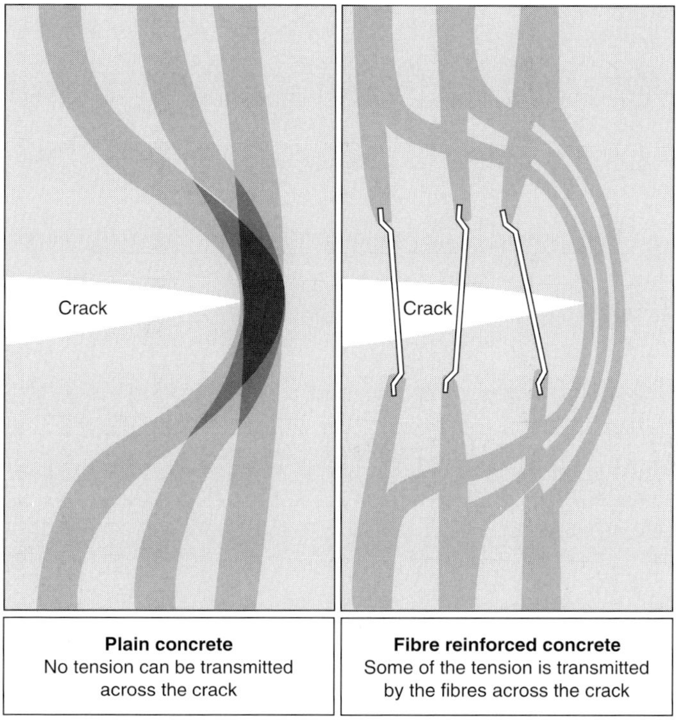

Fig. 1.3. Stress lines in concrete under tension (Tatnall and Kuitenbrouer)[1.16]

which the load–deflection curve deviates from a straight line relationship and is illustrated in Fig. 1.4. Flexural strength is calculated from the load at first crack and the dimensions of the test specimen.

In the Japanese standard, flexural strength is defined in terms of the maximum load and specimen dimensions as the modulus of rupture. This can be seen in Fig. 1.5. The flexural strength of concrete is determined from loading tests on concrete prisms (usually $150 \times 150 \times 450\,\text{mm}^3$) at 28 days. The test is run at a constant deflection rate of 0.5 mm/min. The actual deflection is recorded as a function of load. The test is continued until the deflection is at least 3 mm (1/150th of the span). The area below the curve up to 3 mm deflection is the flexural toughness D_b, expressed in N/mm. The flexural toughness factor or equivalent flexural strength f_e is defined as:

$$f_e = D_b I/(dbh_t^2) = 1/(150)^2 D_b$$

where D_b is the flexural toughness, I is the second moment of area, d is the deflection of slab, b is the distance of applied load from support and h_t is the height of the slab (all in millimetres)

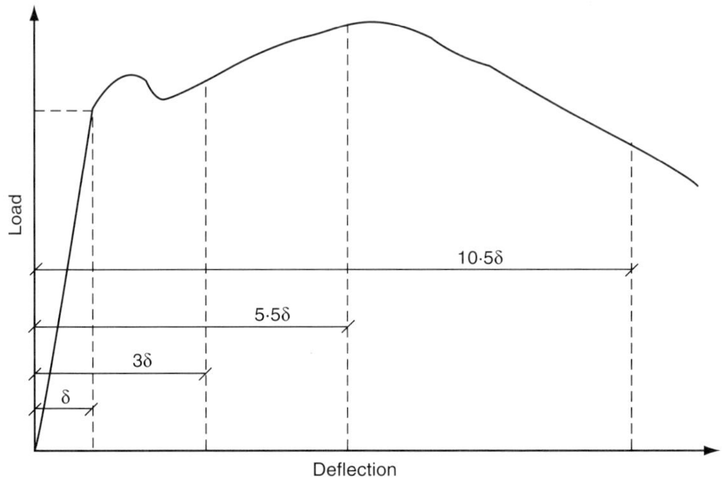

Fig. 1.4. Flexural strength test method used in the United States (δ = deflexion occurring at elastic limit)

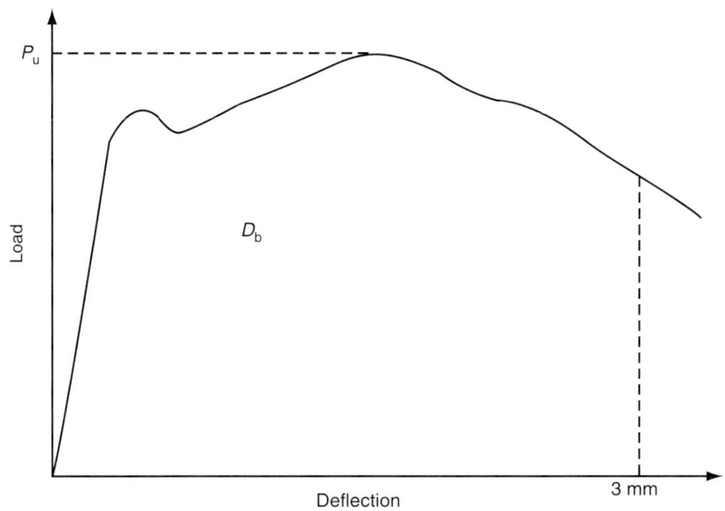

Fig. 1.5. Japanese test method (P_u = ultimate load)

The equivalent flexural strength is the representative value for the reinforcing effect of steel fibres. In Japan, this test is already standard, while in the Netherlands it is included in CUR Recommendation 10[1.17] as a basis for determining the design value of Steel Fibre Reinforced

Table 1.15. Concrete design flexural strengths with steel fibres present

Concrete grade and dosage	Flexural strength: N/mm^2
Plain C30 concrete	2.0
20 kg/m^3 steel fibre C30 concrete	2.8
30 kg/m^3 steel fibre C30 concrete	3.2
40 kg/m^3 steel fibre C30 concrete	3.8
Plain C40 concrete	2.4
20 kg/m^3 steel fibre C40 concrete	3.1
30 kg/m^3 steel fibre C40 concrete	3.6
40 kg/m^3 steel fibre C40 concrete	4.2

Concrete (SFRC). The Dutch design method assumes the mean value of the equivalent flexural strength to be the flexural strength design value f_f.

A manufacturer of anchored steel fibres commissioned TNO, Delft[1.18] to undertake flexural strength tests using fibres embedded in C30 concrete. These tests have resulted in values of mean flexural strength of up to 4.2 N/mm^2, depending on dosage, type and size of fibre. Partly from these results and partly from work undertaken at the UK Cement and Concrete Association the flexural strength values shown in Table 1.15 have been developed. The values are repeated in Table 4.1. The values in Table 1.15 apply to anchored bright wire steel fibres of length 60 mm and wire diameter 1.0 mm.

1.6.8 Post-cracking behaviour

The addition of steel wire fibres to a concrete floor slab ensures that it has load bearing capacity following the appearance of the first cracks. Laboratory tests have resulted in the following theory of the behaviour of steel fibre reinforced slabs. Before the first peak load, a concrete slab exhibits elastic behaviour with the modulus of elasticity being similar to that of non-fibrous concrete. Following the first peak load and with increasing slab deflection, there is a redistribution of the bending moments (stresses) which leads to higher ultimate load capacities and thus enhanced performance.

Calculation methods used to determine the thickness of concrete floor slabs are based upon elastic theory and do not take into account the specific properties of SFRC. This has led to some debate concerning the way in which SFRC floors should be designed. The Netherlands CUR Commission has suggested the adoption of a lower elastic modulus for SFRC to account for the redistribution of stresses and has suggested the use of Japanese flexural strength specifications to account for the additional toughness inherent in SFRC. Section 5.4 shows the results and conclusions from research programmes studying the post-cracking performance of SFRC.

1.6.9 Resistance to impact, fatigue and corrosion

The increased resistance to fatigue provided by SFRC is of particular importance to floors subject to heavy and intensive traffic often found on industrial floors. Such floors must be able to resist the frequent and sudden heavy loads common in industrial areas and therefore the increased impact resistance gained by the use of fibres is an important attribute. Spalling of concrete resulting from the corrosion of steel reinforcement is greatly inhibited when SFRC is specified because of the small diameter of the steel fibres.

1.6.10 Economy

Steel wire manufacturers claim that substantial labour and materials savings can be achieved by the specification of steel wire fibres. Some of the economic advantages are:

- elimination of labour needed for cutting and fixing traditional mesh reinforcement
- quicker levelling of the floor owing to the absence of top reinforcement
- reduction in slab thickness compared with floors designed with plain and mesh reinforced concrete owing to increased flexural strengths
- greater joint spacing.

1.6.11 Specification

For any SFRC application it is recommended that the following be included in the specification.[1.17]

- A description of the desired sub-base (work floor or sheet).
- The required strength class of the concrete.[1.19]
- The required rate of consistency of the SFRC.
- The method of compaction.
- Usage of a plasticizer if required and which type.
- The method of checking homogeneity of the mix.

1.7 Welded steel wire fabric (mesh)

Fabric should comply with the requirements of BS 4483: 1985.[1.20] Information has been used from BS 4483 to compile this section.

1.7.1 Introduction

Steel wire fabric comprises an orthogonal arrangement of longitudinal wires and cross wires welded together at some or all of the cross-over points in a shear resistant manner. The fabric is usually manufactured by machine with the intersection joints formed by electrical resistance welding. Butt welded wires are also often used. The shearing load required to produce failure of a welded intersection should not be less than $0.25Af_y$, where A is the nominal cross-sectional area of the smaller wire at the welded intersection and f_y is the wire's characteristic yield strength. The mesh is usually supplied in bundles bound together in a flat, rolled or folded form.

1.7.2 Quality control

Manufacturers specify wire of grade 460 complying with the relevant British Standards (BS 4449, BS 4461, BS 4482) to produce the fabric. The number of broken welds must not exceed 4% of the total and must not exceed half the number of cross welded joints along any one wire.

1.7.3 Dimensioning

Steel wire fabric is available in the wire diameter and spacing arrangements shown in Table 1.16.

1.7.4 General

Steel wire fabric is assumed to carry the tensile force developed in the concrete owing to the contraction of the slab due to shrinkage and expansion. Consequently mesh allows greater joint spacings. Fabric is often used in long strip floor construction as it can be placed conveniently without the need for cutting. Where necessary, fabric sheets should be lapped at their edges and ends by 450 mm. Overlapping can result in an unacceptable build up of thickness of reinforcement. When using wire guided vehicles, interference with control signals needs to be considered (i.e. careful placing of the mesh at a specific depth within the slab may be needed).

1.7.5 Polymer grid reinforcement

Although they do not contribute to the flexural strength of concrete, polymer grids are a recent development that may provide an economical

Table 1.16. Preferred range of designated fabric types and stock sheet size[1.20]

Fabric reference[a]	Longitudinal wires			Cross wires			
	Nominal size: mm	Pitch: mm[b]	Area: mm^2/m	Nominal wire size: mm	Pitch: mm	Area: mm^2/m	Mass: kg/m^2
Square mesh							
A393	10	200	393	10	200	393	6.16
A252	8	200	252	8	200	252	3.95
A193	7	200	193	7	200	193	3.02
A142	6	200	142	6	200	142	2.22
A98	5	200	98	5	200	98	1.54
Structural mesh							
B1131	12	100	1131	8	200	252	10.9
B785	10	100	785	8	200	252	8.14
B503	8	100	503	8	200	252	5.93
B385	7	100	385	7	200	193	4.53
B283	6	100	283	7	200	193	3.73
B196	5	100	196	7	200	193	3.05
Long mesh							
C785	10	100	785	6	400	70.8	6.72
C636	9	100	636	6	400	70.8	5.55
C503	8	100	503	5	400	49	4.34
C385	7	100	385	5	400	49	3.41
C283	6	100	283	5	400	49	2.61
Wrapping mesh[c]							
D98	5	200	98	5	200	98	1.54
D49	2.5	100	49	2.5	100	49	0.77
Stock sheet size[d]	Longitudinal wires			Cross wires			Sheet area
	Length 4.8 m			Width 2.4 m[e]			11.52 m^2

[a] When specifying a steel wire, fabric reference codes should be used. Reference letters A, B, C and D represent square, structural, long and wrapping mesh respectively. The numbers in the reference represent the area of steel of the longitudinal wires per metre width of fabric.
[b] Pitch is defined as the centre to centre spacing of wires in a sheet of fabric.
[c] Wrapping mesh: wire usually of grade 250 for use in wrapping fabric.
[d] Stock sheet size: fabric types A and B are delivered in standard sheets of $4.8 \times 2.4\,m^2$, or in scheduled size sheets. Fabric type C is available in sheets or rolls.
[e] When specifying fabric sheets the width is the overall dimension measured in the direction of the cross wires.

alternative to steel fabric. The inclusion of a polymer grid aids the early age crack control of concrete in a similar manner to fibre reinforced concrete (see Section 1.5.6). The lightweight, non-rusting and non-magnetic nature of such grids may prove beneficial. Additional research is needed before polymer grid reinforcement can be specified commonly.

1.8 Reinforcing bars (rebar)

Steel reinforcing bars should comply with BS 4461: 1978 (1984).[1.21] Information has been taken from BS 4461 to produce this section. Steel reinforcing bars (rebars) are delivered to site either in stock lengths, scheduled lengths or are cut and bent to specified shapes. Stock lengths are usually 6 m or 12 m. Bars delivered to site should be tagged with an identification number or code relating to the floor's reinforcing schedule or design drawings. The preferred diameters are shown in Table 1.17.

High yield grade 460 steel is used for concrete floor slabs (characteristic tensile strength is 460 N/mm^2). Bars should be free from defects and, in order to ensure correct bonding to the concrete, there should be no loose rust, scale, grease or dirt present when the concrete is cast. Also if wire guided vehicles are to be used, rebars should be set at a sufficient depth to avoid interference with control signals. The inclusion of steel reinforcement bars in concrete floors permits thinner slabs to be designed.

1.9 Surface finishes

1.9.1 Introduction

When selecting a specific type and method of finish a number of factors need to be considered. These factors include the type of traffic and loading the floor will encounter, and the need for abrasion, impact and chemical resistance. With the development of many specialist toppings and surface treatments designers are able to specify lower strength concrete for the floor slab and provide the surfacing requirements in the topping material. A number of materials are available which can improve the wear resistance, chemical resistance and the general appearance of a concrete floor as well as reduce its slipperiness and susceptibility to dusting.

1.9.2 Finishing techniques

To ensure good performance from a concrete floor surface the concrete

Table 1.17. Preferred sizes of rebar[1.21]

Nominal diameter: mm	8	10	12	16	20	25	32	40

mix must undergo full compaction. Surface water often results from vibration actions performed on the concrete surface and should be removed if a durable, wear resistant surface is to be achieved.

1.9.3 Cement based toppings

It can be very expensive to construct a floor of very high strength concrete. If a floor is susceptible to particularly abrasive conditions, it may be more cost effective to use a lower strength concrete with a high strength topping. Toppings are thin layers of cement-rich, fine aggregate concrete with a high shrinkage potential. High strength toppings should have an aggregate/cement ratio of about 3:1 and consist of about 30% good quality concreting sand with 10 mm single-sized coarse aggregate (often crushed rock). Debonding, curling and cracking are problems which have to be minimized with careful consideration to be taken in the design and construction of these floors. Toppings are applied in one of the following ways.[1.22]

(i) *Monolithic construction.* The topping is applied when the concrete is still in a plastic state. This allows the topping to become structurally integral with the slab. High strength toppings are usually between 12 and 20 mm thick. This form of construction eliminates the risk of the topping debonding from the slab.

(ii) *Bonded construction.* The topping is bonded to the slab after it has hardened. It has little structural value and is therefore not included in the structural depth of the slab. To apply the topping, the slab's coarse aggregate must be exposed by the use of mechanized plant, such as a pneumatic scabbler or shot blaster. The surface should be rigorously cleaned of dust before the topping is applied. Bonding agents or admixtures are often used.

(iii) *Unbonded construction.* Requires increased topping thickness and is constructed above a damp-proof or isolating membrane interposed between the topping and the slab. This method is often used when re-surfacing is needed.

1.9.4 Curing compounds

Curing is a vital operation in the production of a hard wearing concrete floor surface. The main objective is to prevent early drying out of the surface and therefore to allow full hydration to take place, resulting in a greater final strength and abrasion resistant concrete. Taking good care in curing reduces the risk of plastic cracking, dusting and drying shrinkage. Although traditional curing methods such as the application of wet hessian sacks and polythene sheets are still often used, it is now common to specify a curing agent, often in the form of an acrylic polymer solution which impregnates the concrete surface forming a membrane. Application of the curing membrane is usually in the form of a spray onto the newly laid concrete after the moisture sheen has evaporated, or after final powerfloating. Many curing compounds combine the two functions of curing and sealing thereby resulting in increased concrete surface protection.

Abrasion resistance can be greatly enhanced by the application of a resin based sprayed-on membrane. Polyethylene sheet proved to be the next most effective, followed by the wet hessian sack method and finally air alone. It was reported that the need for a good curing technique was particularly important when a high water/cement ratio was used.

1.9.5 Dry shake floor hardener (sprinkle finishes)

The surface of a concrete floor slab may be enhanced by the application of a metallic, non-metallic or natural (quartz) dry shake floor hardener. Ready-to-use pre-blended materials consist of selected aggregates mixed with cement. When sprinkled and trowelled into the fresh wet surface of concrete floors (monolithic construction) they form a dense, toughened, wear resistant surface. The use of a dry shake topping can also prove beneficial in providing a smooth, anti-slip and non-dusting surface with increased resistance to the penetration of oils and greases. It is becoming common to apply the topping with an approved automatic spreader used in conjunction with a laser screeder, with the full specified quantity applied evenly onto the concrete immediately following screeding. Tests[1.23] using an accelerated wear abrasion machine have demonstrated how dry shake floor hardeners increase resistance to abrasion. The test results are summarized in Table 1.18.

Table 1.18. Increasing abrasion resistance with dry shake hardeners

	Abrasion depth: mm
40 N/mm^2 repeat powerfloated concrete[a]	0.4
Non-metallic dry shake floor hardener	0.05
Metallic dry shake floor hardener	0.02

[a] The control specimen displayed the best performance which can be realistically achieved from an untreated concrete surface. A curing agent and sealing agent are often applied after finishing with the dry shake topping.

1.9.6 Liquid surface treatments

Several surface treatments to improve the properties of the surface of a floor are available. Prior to application the floor surface quality should be good, otherwise the benefit gained may only be temporary. Chemical hardeners such as sodium silicate and magnesium fluorosilicate are often used in solution to increase wear resistance and prevent dusting. The surface to which the treatment is to be applied should be dry and clean to allow the hardener to react chemically with the concrete to produce a case hardening effect. Most hardeners are spray applied approximately two weeks after concreting.

There are many coatings available in the form of sealers and paints. Solvent based resin sealers have been proven to improve the abrasion and chemical resistance of a concrete surface significantly, so improving its performance and maintenance costs. The sealer will also provide the floor with a high resistance to the penetration of oils, cleaning detergents and many other harmful chemicals. When pigmented, a sealer can also improve the appearance of a concrete floor.

Tests[1.23] have been performed to estimate the significance of liquid surface treatments. It was concluded that penetrative resin seals were found to produce a significant increase in the abrasion resistance of a concrete surface and that chemical hardeners had much less effect.

1.9.7 Basic recommendations

The following recommendations (Reproduced from TR34)[2.1] should be considered when specifying finishing treatments:

(i) *In the majority of industrial environments a satisfactory degree of abrasion resistance can be achieved with repeated power trowelling and effective curing without modifying the surfaces of the floor in any other way.*

(ii) *In a light industrial environment abrasion resistance can be achieved with a single pass of the power trowel combined with effective curing.*

(iii) *In a heavy industrial environment some modification of the surface is required prior to or after power trowelling and curing, although the most effective way by which it can be achieved is still to be ascertained. This may include the application of a 'sprinkle' finish, high strength topping, or penetrating in surface seal.*

2. Construction

2.1 Introduction

A correctly designed and constructed concrete floor combines the advantages of hard wear, long life and the ability to carry heavy loads at low costs. The purpose of a floor will vary according to application and each floor requires its own individual characteristics including strength, abrasion resistance, flatness and aesthetics. An important factor commonly taken into consideration is speed of construction and the savings which accrue from fast-track flooring.

2.2 Principal flooring issues

Four important issues to be considered in industrial concrete floors are:

- irregular floor tolerances
- cracking
- joint breakdown
- lack of abrasion resistance

One of the unique features of ground floors is the relationship between design, construction and performance. All of the above issues are influenced by both design and construction. The following sections provide construction guidance which will maximize performance and minimize the possibility of development of the above situations.

2.3 Traditional construction methods[2.1]

2.3.1 Long strip construction

Long strip floor construction has been established as the conventional method of construction since the early 1970s. The floor is laid in a series of strips up to 6 m wide using timber or steel formwork. Every second strip is concreted initially, leaving infill unconcreted strips. The infill strips are concreted several days later using the originally laid strips as the formwork. Strips can be up to 60 m or more in length. Floor slabs are often thicker than 150 mm and steel mesh reinforcement is usually provided. The floor may be placed and compacted in two layers, the upper layer being placed whilst the lower is still plastic. Compaction takes place using either internal poker vibrators or twin-beam vibrating compactors running on the formwork or on the previously cast slabs.

2.3.2 Wide strip construction

Wide strip construction developed in the United Kingdom following the introduction of a wide span compacting beam known as the 'Razorback' (space frame compacting beam, see Fig. 2.1) developed in the US. Razorbacks enable floors to be laid in strips of width up to 25 m although it is more common to work in the range 9–15 m. Laying the floor follows the same principles as for the long strip method. The concrete is placed between formwork, levelled, screeded, compacted and left to cure. Often, the use of two-layer construction to incorporate mesh reinforcement is uneconomical so stools or large diameter circular fabric supports are used to enable the mesh to be positioned prior to the placing of the concrete.

The advantage of wide span construction over long strip construction is that a greater daily output of completed floor area can be achieved with a similar sized labour force. As larger areas can be constructed, more skilled finishers are required and often multi-headed powerfloats are used enabling one-layer compaction.

2.4 Large bay construction[2.1, 2.2]

Concrete floors are frequently laid using large bay or large pour methods so as to increase the daily output and to speed up the project. The first UK method was developed by A. Monk & Co. Limited working with Silidur SA of Belgium and differed from conventional ground floor construction in that no side forms were required, except to contain a day's pour. High

Fig. 2.1. Use of 'Razorback' enables greater widths of concrete to be placed

workability concrete has a slump in excess of 150 mm which is made possible by the addition of super-plasticizers. The high slump value allows concrete to be poured directly from a truck mixer and to be spread manually. The concrete almost self levels such that a satisfactory level can be achieved by undertaking final adjustments based upon levels provided by laser transmitters. Any discrepancies from true level can be corrected using timber screed boards to bring the surface to the correct level. Compaction is achieved by lightweight screed beams or vibrating pokers. The principal disadvantages of the method are segregation of the aggregate with the larger particles sometimes sinking to the bottom of the slab and the poor surface regularity often achieved. Segregation can lead to a high concentration of fines at the surface and consequent loss of abrasion resistance. For this reason, such floors are often treated with abrasion enhancing toppings (e.g. Amorex).[2.3] This technique allows the use of steel mesh reinforcement which is usually laid out a day ahead of the placing of the concrete.

2.5 Laser screed construction[2.2, 2.3]

Hughes Group and John Kelly (Lasers) Ltd[2.3] introduced laser guided screeding machines to Europe and Scandinavia in the 1980s. Precision Concrete Floors Ltd (PCF),[2.3] who had been constructing concrete floors using traditional methods, were the first UK company to adopt laser screeding. By the mid 1990s, there were six laser guided screeding machines constructing concrete floors in the UK, one of which is now described.

2.5.1 *The Somero S240 Laser Screed™* [2.2]

The machine described in this section is one of several types of laser guided screeding machines. The Somero S240 Laser Screed™, one of several such machines now in service, has been described as *'the most significant advance in the construction of concrete floors in the last decade'*. Laser guided screeding machines combine state-of-the-art laser control systems with conventional mechanical screed mechanisms. The machines have four wheel drive, four wheel steer including 'crab' steer for awkward areas and are operated by one person seated at a point of maximum visibility (see Fig. 2.2).

Mounted on the twin axles a circular fully slewing turntable carries a counter balanced telescopic boom, typically having a 6 m reach on the end of which is attached a 3–4 m wide screed carriage assembly which comprises a plough, an auger to spread the concrete accurately and a vibrating beam for compaction. Test results have shown that such machines can compact concrete to depths in excess of 300 mm.

Fig. 2.2. Somero S240 Laser Screed™

A self levelling laser transmitter is fixed at a visible point close to the work so as to project a 360° rotating beam across the working area. Depending on the type of transmitter, various inclinations of floors can be achieved including level, single and dual grades. The level of the laser screeder is controlled by a laser beam which activates receivers mounted on the screed carriage assembly. During concreting the signals are relayed continuously to an on-board control box which automatically controls the level of the working screed head by direct intervention on the machine's hydraulic system. The laser transmitter rotates at 300 rpm so that the height of the screed carriage is adjusted five times per second.

2.5.2 *Laser screed operation*

Laser guided screeding machines are used in conjunction with mixer trucks which place concrete 25–35 mm above the finished floor level. The positioning of the laser screeding machine, mixer trucks, slip membrane, joints and reinforcement requires careful organization so as to attain maximum output. Once the concrete has been deposited from the mixer, the boom of the screeding machine is extended over the fresh pour and the screed carriage is lowered until the receivers lock onto the signal generated by the laser transmitter. The boom is then retracted, drawing the screed carriage towards the operator across the concrete placing, compacting and screeding simultaneously.

In one pass a laser screeding machine can place compact and screed 20 m^3 concrete in under two minutes. Because of the geometry of the horizontal auger, screeding takes place from left to right with an overlap between sequential screeding runs of 300 mm so as to ensure optimum level and surface regularity across the entire floor slab. As the pour is contained by the building perimeter no formwork is required except to allow for doorways and drains or to contain a day's pour for floors which cannot be finished in one day.

Output depends on topological details, type of reinforcement used (fabric can pose several problems as laser screeders tend to lift the mesh out of the concrete) and the speed at which the concrete supplier can deliver the concrete to the site. Outputs of between 2000 m^2/day and 4000 m^2/day are normal and 5000 m^2/day has been reported.

2.5.3 *Laser screeding level control*

The screeding level of the machine's screed carriage is maintained by an automatic laser controlled system. Laser receivers mounted at each end of the screed carriage detect the reference datum plane emitted by a laser level transmitter situated near the work area. The on-board control box checks and adjusts the screed carriage level in relation to the laser plane five times per second.

2.5.4 *Advantages associated with laser screeding machines*

Contractors using laser screeding machines have reported the following advantages over traditional long strip construction.

- Higher strength, denser and more durable floors.
- Wide bay construction with maintained tolerances.
- Flatter floors (see Fig. 2.3).
- Working method ensures high productivity as it eliminates manual screeding.
- Ensures construction programmes are kept to time, enabling possibilities of earlier use of facilities.
- Damage-prone construction joints are kept to a minimum.
- Choice of mesh or fibre reinforcement.
- Larger areas of floor can be placed, screeded, vibrated, compacted and left to cure in a single day. A floor area of 5000 m^2/day has been reported but outputs of 2000–4000 m^2/day are more common.

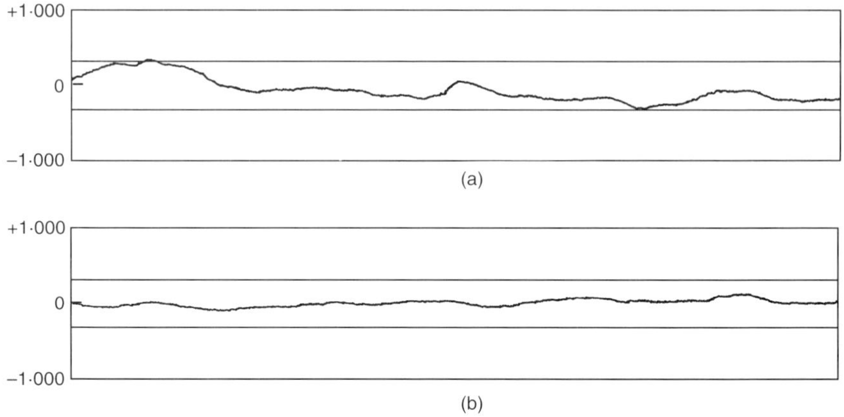

Fig. 2.3. Comparison of floor profiles achieved with laser screed (lower trace) and manual screeding (upper trace)

2.6 Finishing and curing processes

Once the floor has been laid and screeded the final stages of finishing and curing are important in order to achieve a flat smooth durable floor.

2.6.1 Floor finishing

Finishing is a critical and skilful operation which takes place after the floor has been screeded. Preliminary finishing involves the use of a lightweight alloy straight edge which is drawn across a freshly laid surface to produce a flatter surface than that when screeded. Once the floor has gained sufficient strength to support a person, powerfloating takes place using walk-behind mono-powerfloats or multi-headed ride-on powertrowels in order to achieve a hard, dense and flat surface. On floors where high surface tolerances are required such as in warehouse racking buildings, a combination of powertrowels and straight edges may be used. Once finishing has taken place the concrete is left to cure.

2.6.2 Curing

Curing is an essential part of the construction process. Some consider that the labour intensive traditional method of covering the floor with wet hessian for at least seven days is the most effective curing method. If water escapes too early the upper part of the slab will dry out too quickly with slower rates of curing occurring below. This can lead to low abrasion resistance and slab curling. Curing time and hence concrete strength is affected by temperature, wind, rain and sun. Shading may be sufficient to allow the concrete to cure. Alternatively, a spray cure membrane such as

Proseal, manufactured by Armorex,[2.3] may be applied immediately following powerfloating.

2.7 Slip membranes[2.1, 2.2]

2.7.1 The purpose of slip membranes

The purpose of slip membranes is to reduce the coefficient of friction between a concrete slab and its sub-base. It is not intended to act as a damp-proof membrane (DPM) although the material does allow the retention of some moisture. If a DPM were to be used then no moisture would escape downwards from the slab. Perforations cannot be entirely avoided and so an amount of moisture may escape into the sub-base. The slip layer reduces the coefficient of friction so allowing the concrete to move more easily as it shrinks during the curing process, hence reducing stresses in the slab, decreasing the possibility of cracking and assisting in the development of induced joints.

2.7.2 Slip membrane materials

At present there is no British Standard governing the use and type of slip membrane although polyethylene sheets are commonly used with thicknesses of 1000 gauge (250 microns) and 1200 gauge (300 microns).

2.7.3 Importance of sub-base with regard to slip membranes

Slip membranes are installed between the slab and sub-base immediately before the concrete is placed. They have to be strong enough to withstand construction traffic such as truck mixers. It is important to overlap the sheets by at least 200 mm and use tape to ensure adjacent sheets do not move or separate. Care must be taken to avoid wrinkles and rips which may induce cracks in the strengthening slab.

If ruts are present in the sub-base surface and the slip membrane sheets are placed over them, voids may form beneath the slab when the concrete is placed. Weak spots will develop, leading to the possibility of cracking when load is applied. If the cross-section of the sub-base consists of rises and falls or peaks and troughs, the slip membrane might fail to do its task as the shape of the sub-base may cause interlock and therefore limit the horizontal movement of the slab as it shrinks. Stresses could increase and cracking ensue. It is important to ensure that the sub-base is as flat as possible and that care is taken in laying the slip membrane sheets.

2.7.4 Use of slip membranes in post-tensioned slabs

A post-tensioned slab has tendons running through its length. Once the concrete has gained enough strength, the tendons are tensioned, so compressing the slab and increasing the strength of the floor by preventing tensile stress from developing in the concrete. Two layers of slip membrane can be used to reduce the coefficient of friction between the slab and the sub-base. This may lead to a reduction in the amount of reinforcement required within the slab and also to a reduction in the number of contraction joints. When two layers of slip membrane are used for their slip characteristics as opposed to their strength, the thickness may be reduced. Typically, a sheet used in two-layer construction might be of 500 gauge (125 microns). Greater care must be taken in the handling and placing of thin membrane sheets.

2.7.5 Considerations in the design and use of slip membranes

There are no specific guidelines regarding the use of slip membranes and the designer has the choice of whether or not to use a slip membrane. The choice will be based on considerations of shrinkage, strength, curling and joint details. The purpose of the slip membrane is to allow the concrete slab to shrink freely and so reduce the levels of stress developed by restraint to movement. Allowing movement reduces the number of contraction joints required, but the joints which are provided may be more active. A consequence of eliminating the slip membrane is the development of interlock between the sub-base and the slab. Slab/sub-base friction is enhanced and the coefficient of friction increases from 0.2 to 0.7 so inhibiting the movement of the slab and requiring more contraction joints, each moving by less. For this reason, omitting the slip membrane might be beneficial in warehouse racking buildings where large joint movements could render a floor unsuitable for pallet and fork lift trucks owing to the possibility of such equipment becoming unsteady when carrying load across joints.

Curling is caused by differential moisture loss between the surface and base of a concrete slab. When a slip membrane is used, the majority of moisture escapes from the slab through its upper surface which then shrinks and cures faster resulting in the slab attempting to curl up at its edges. Eliminating a slip membrane allows moisture to escape through the base of the slab which leads to more even curing of the slab and so reduces curling. If too much water escapes through the base of the slab, instead of curling upwards, the slab may develop hogging with the edges of a slab attempting to curl downwards. Blinding (i.e. providing fine material at the surface) the sub-base can reduce hogging by allowing moisture to escape from the slab whilst at the same time preventing it from flowing away through the sub-base. The ideal slip membrane would:

- allow the slab to move relative to the sub-base as it shrinks during curing, enabling fewer joints to be constructed;

- be easy to handle and strong enough not to tear;
- be perforated so as to allow a controlled amount of moisture to escape and so eliminate curling.

2.8 Joint details

In the early life of a concrete floor the slab will contract and be in danger of cracking. This can be controlled by an arrangement of joints. In floors where operating conditions permit the presence of cracks, for example in cases where hygiene and dust control are of secondary importance, the number of joints can be reduced and additional reinforcement provided. Particular attention should be paid to the alignment, setting and compaction of concrete at joints.

2.8.1 Movement joints

Movement joints are provided to ensure minimum restraint to movement caused by moisture and thermal changes in the slab. Movement joints are designed to allow the slab to contract. Expansion joints are used in regions where temperature changes can be substantial over a short period of time and are rare in the UK.

2.8.1.1 Formed Doweled Contraction (FDC) joint. This type of joint (Fig. 2.4) is provided at the ends and sides of a construction bay. It includes debonded dowel bars. To ensure the dowel bar and sleeve arrangement can move the bars are debonded using one of three techniques illustrated in Fig. 2.5.

2.8.1.2 Induced Doweled Contraction (IDC) joint. This type of joint is designed to allow movements as the concrete shrinks. It is used in construction bays which are designated large (any area greater than

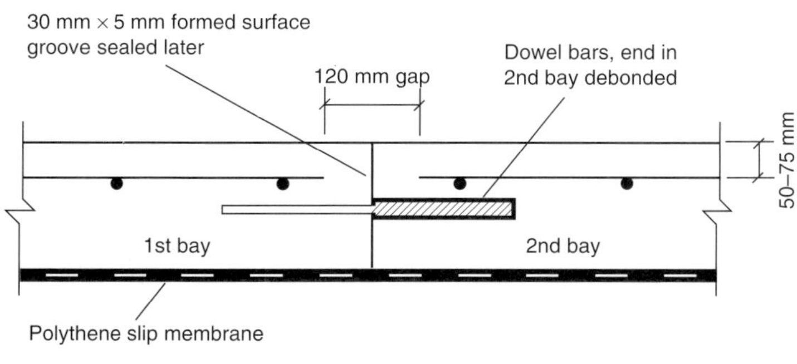

Fig. 2.4. A Formed Doweled Contraction (FDC) joint

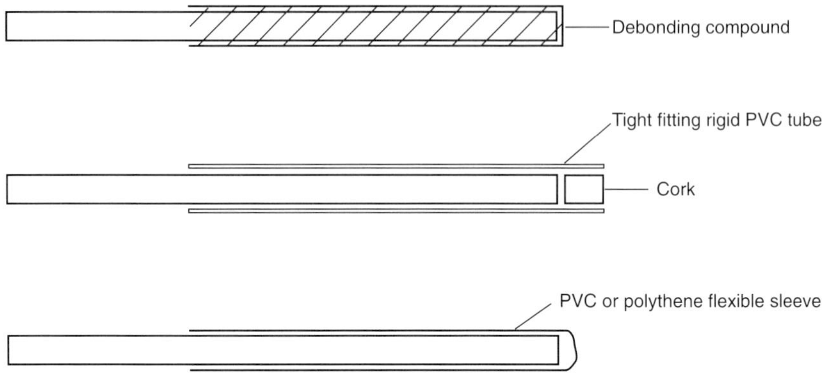

Fig. 2.5. Three methods of debonding doweled joints

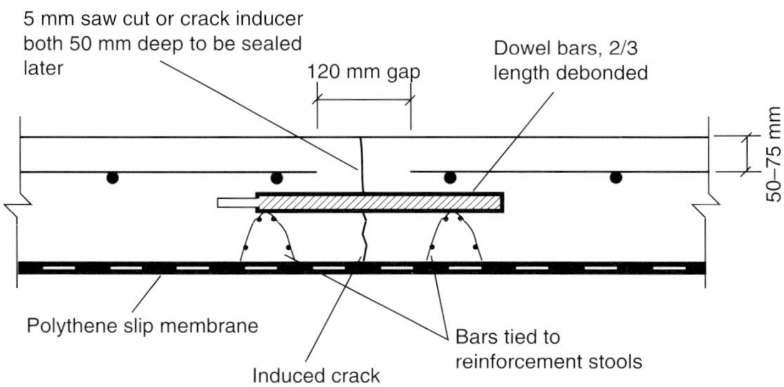

Fig. 2.6. Induced Doweled Contraction (IDC) joint

1000 m^2) and which have been poured in a continuous operation. This type of joint (Fig. 2.6) is formed by crack induction (Section 2.8.3). The dowel bars enable vertical load transfer and are fixed prior to the placing of the concrete.

2.8.1.3 Induced Contraction (IC) joint. Figure 2.7 shows the type of IC joint used in large construction bays. It reduces the cost of joints by eliminating dowel bars and reduces the risk of joint breakdown owing to poor workmanship in the placing of the dowel bars before concreting. Load transfer depends on aggregate interlock across the induction groove. Floors incorporating this type of joint are suitable for lightly loaded applications where reinforcement requirements are minimal.

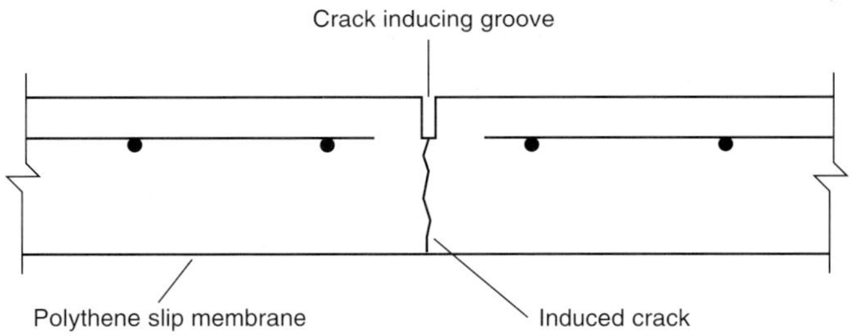

Fig. 2.7. Induced Contraction (IC) joint

2.8.1.4 Isolation joints. Isolation joints or full movement joints are used to permit movements around the fixed parts of buildings such as recesses, walls (Fig. 2.8), drains and columns. They keep the concrete slab separate from fixed elements. At columns there are two different methods of isolation, diamond or circular (Fig. 2.9). The circular joint is often preferred as it uses less expensive formers and has no stress inducing sharp corners.

2.8.2 Tied joints

Tied joints hold two construction bays together using tie bars and are primarily used to restrain movement and contraction in the horizontal plane. The tie bars also act as a form of stress relief as they are bonded within the concrete and provide a large amount of strength across the

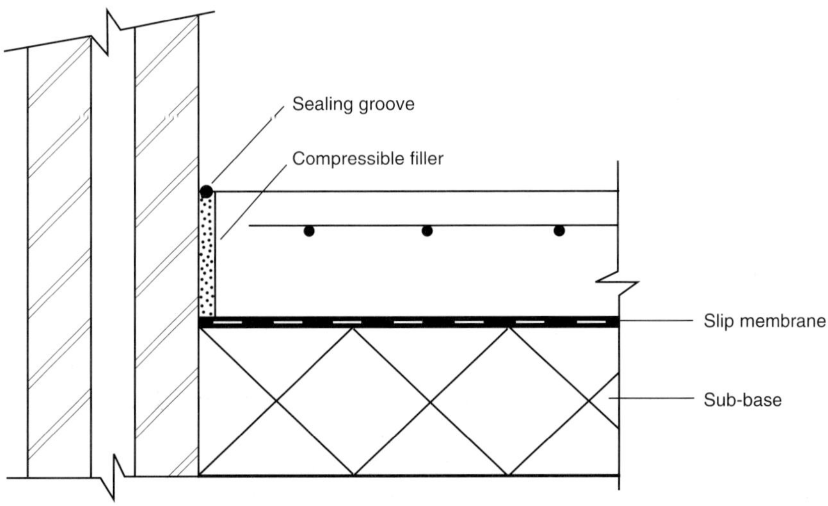

Fig. 2.8. Isolation joint at a wall

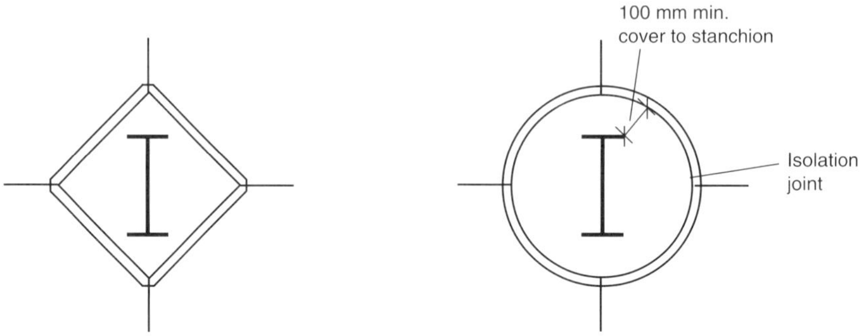

Fig. 2.9. Circular and diamond isolation joints at stanchions (plan view)

joints. Tie bars are inserted in pre-drilled holes in the formwork as the concrete is placed.

2.8.2.1 Formed Tied (FT) joint. This type of tied joint (Fig. 2.10) is provided around the edges of a construction bay or at a stop-end commonly when constructing concrete floors using the long strip method and has proved successful in controlling movement at joints. A groove is provided to allow the joint to be sealed (Section 2.8.5) and is formed by a strip placed on the edge of the first bay cast.

2.8.2.2 Induced Tied (IT) joint. Long strip and large bay pours are prone to cracking. Using induced tied joints (Fig. 2.11) construction bays are reduced in size to reduce the possibility of cracking. As the floor attempts to shrink it is restrained by the ties. Cracking may occur and a crack inducer is placed in the surface of the slab encouraging the slab to crack at the joint.

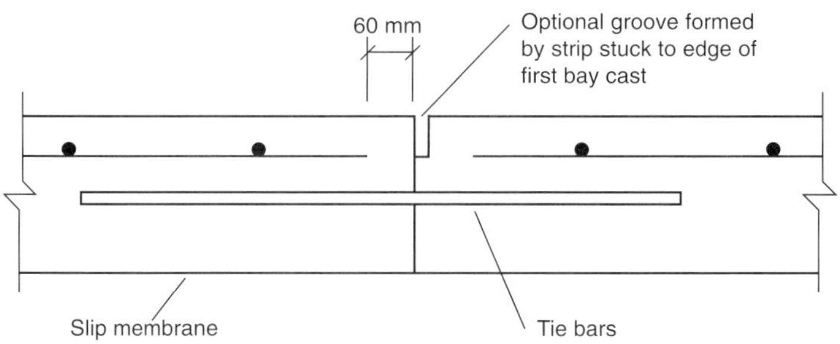

Fig. 2.10. Formed Tied (FT) joint

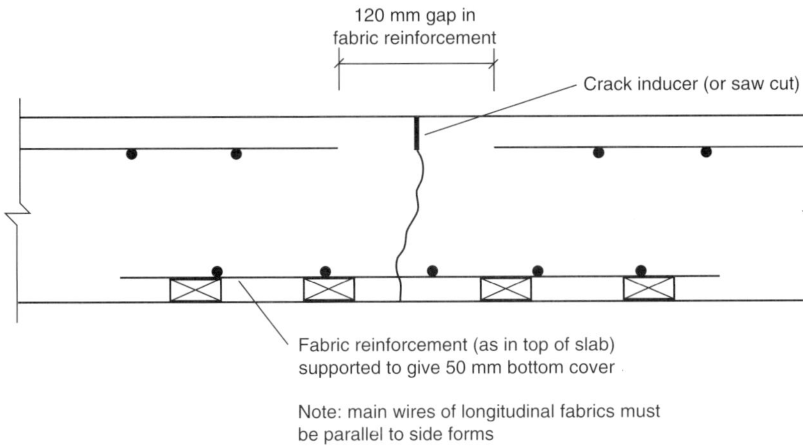

Fig. 2.11. Induced Tied (IT) joint

2.8.3 Crack induction methods

2.8.3.1 Sawn joints. The saw cut acts as a line of weakness which is incorporated into the slab at the position of the joint such that the slab will crack at that point owing to an increase in the tensile stress in the remaining depth of slab. Saw cuts are formed when the concrete has gained sufficient strength to withstand the effects of a concrete saw but not so much that the effect of sawing would damage the floor. Sawn joints are particularly durable. They are expensive and can cost up to ten times as much as wet formed joints. The saw cut breaks the upper layer of reinforcement and the groove has a depth of 40 mm or 50 mm (Fig. 2.12).

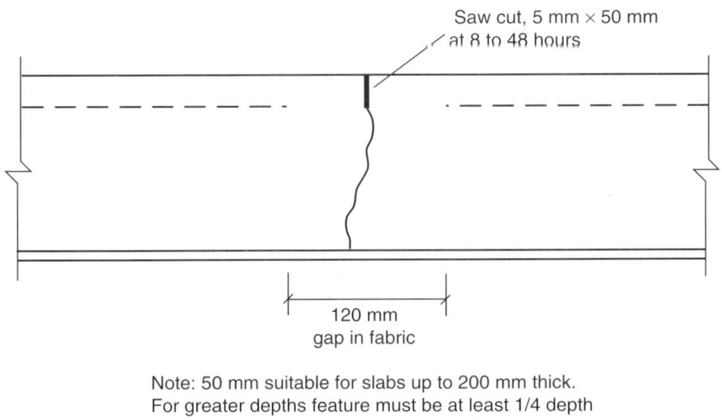

Fig. 2.12. Crack induction joints using the saw cut technique

As well as the joint being the most durable crack induction form, a sawn joint is also very serviceable with no difference in level at each side of the cut. This aids flatness which is important on heavily trafficked floors.

2.8.3.2 The timing of forming sawn joints. Joints should be cut when the concrete has gained sufficient strength to support the weight of the cutting equipment but before it has gained sufficient strength that sawing might loosen or pull out aggregates or fibre reinforcements. A suggested sawing time scale is between 24 and 48 h after initial concrete set. This leaves a time window, as shown Fig. 2.13 assuming the following.

- Concrete is mixed 1 h before placing.
- First concrete is mixed at 7.00 a.m.
- Last concrete is mixed at 4.00 p.m.
- Concrete is placed between 8.00 a.m. and 5.00 p.m. on day 1.
- Assume concrete takes 6 h to reach initial set.
- Therefore, earliest initial set = 1.00 p.m.; latest initial set = 10.00 p.m.

If the concrete is to be sawn between 24 and 48 h after the initial set, the first cut can be performed 24 h after the latest initial set so as to ensure that all of the concrete has gained sufficient strength. All saw cutting must

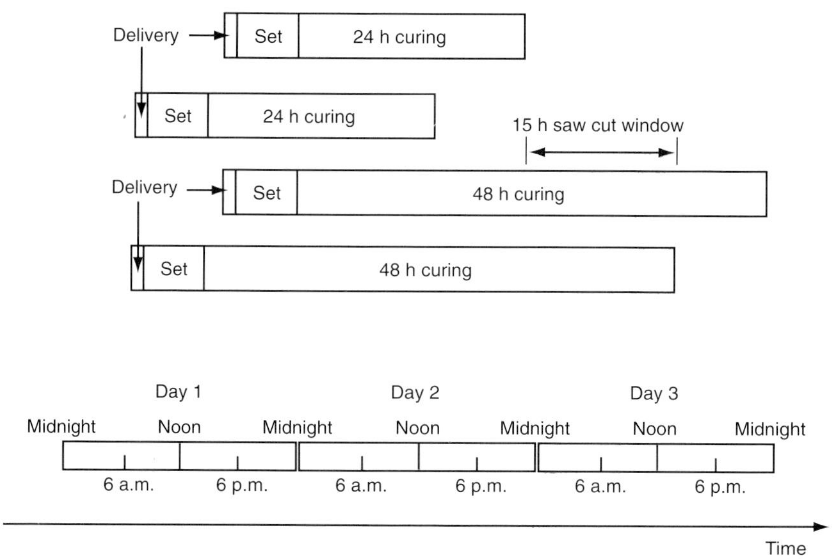

Fig. 2.13. The timing of forming sawn joints

be finished 48 h after the earliest initial set. In this example, saw cutting can commence at 10.00 p.m. on day 2 and must be finished by 1.00 p.m. on day 3. This gives 15 h sawing time.

2.8.3.3 'Soff-cut' sawn joints.[2.4] Soff-cut is a way of forming crack induction joints within a slab prior to the concrete developing initial set. The saw cut is made immediately the powerfloats or powertrowels have finished, saving time and forming an economical joint. The Soff-cut system is used on US highway pavements and a Soff-cut joint serves the same purpose as the more conventional sawn joint.

2.8.3.4 Other methods of crack induction. Figure 2.14 shows traditional crack inducers involving the use of metal or plastic strips which are inserted into the wet concrete after placing. The upper section of the strip is removed once the concrete has hardened leaving the lower section in the slab to form the crack inducer. The metal insert (Fig. 2.14(*b*)) is shaped so as to not disrupt the concrete surface when inserted. It is strong enough to be able to be pushed into the concrete vertically. The zip strip is used more commonly in the US and comprises a plastic extrusion made

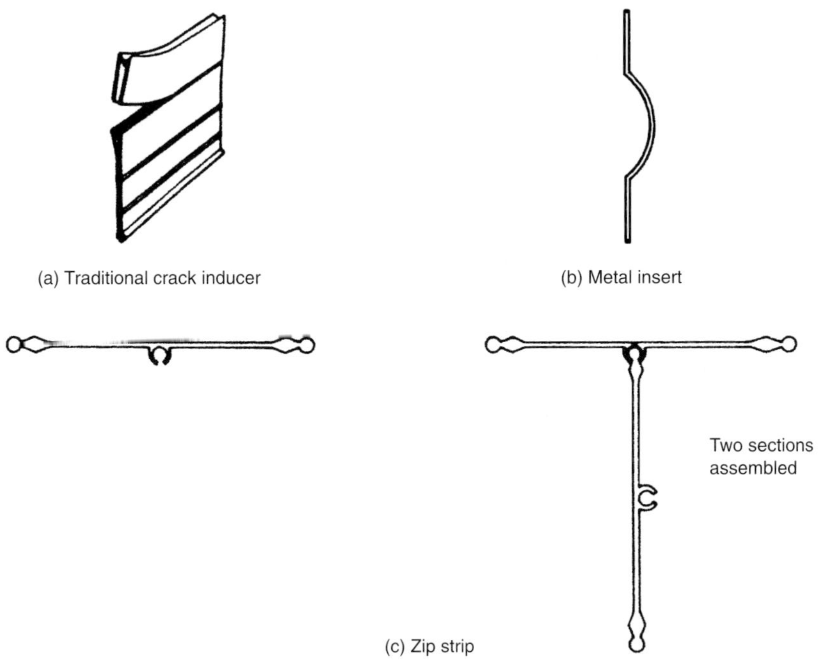

Fig. 2.14. Traditional crack inducers

from two identical parts to form a T-shape. The rigidity of the plastic section allows the vertical section to be pushed into the concrete and the top section is removed for reuse prior to powerfloating. Immediately following concreting a groove is formed using a bricklayers' trowel against a string line to ensure accuracy. Once the inducing strip is in place further compaction and vibration is necessary to ensure that no air has been introduced into the concrete.

2.8.4 Position of joints

Table 2.1 shows joint spacings for various types of concrete. The joint spacings in the table have been used successfully for many years in the UK. Some consider that spacings can be greater than 12 m for steel fibre reinforced floors. Whilst 14 m joint spacings will probably be acceptable for most applications, the additional movement which would occur at joints might lead to loss of aggregate interlock and joint degradation. Joint spacings are frequently designed to coincide with column spacings or other features such as recesses or floor width changes. In any project it is necessary to develop a joint layout identifying all practical issues prior to the placing of concrete.

Table 2.1. Common concretes and suggested joint spacings

Concrete type	Joint spacing: m
Plain C30 concrete	6
Micro-silica C30 concrete	6
20 kg/m^3 ZC 60/1.00 steel fibre reinforcement C30 concrete	6
30 kg/m^3 ZC 60/1.00 steel fibre reinforcement C30 concrete	8
40 kg/m^3 ZC 60/1.00 steel fibre reinforcement C30 concrete	10
Plain C40 concrete	6
Micro-silica C40 concrete	6
20 kg/m^3 ZC 60/1.00 steel fibre reinforcement C40 concrete	6
20 kg/m^3 ZC 60/1.00 steel fibre reinforcement C40 concrete	10
20 kg/m^3 ZC 60/1.00 steel fibre reinforcement C40 concrete	12

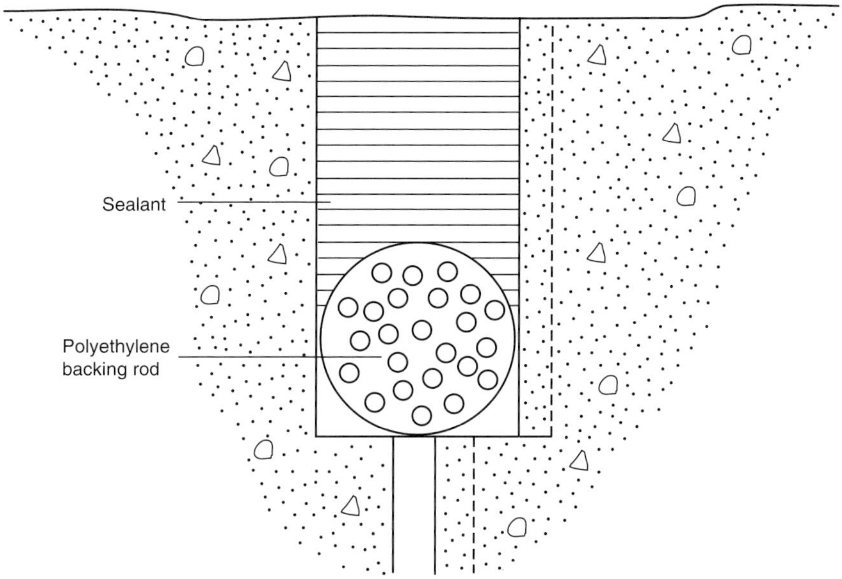

Fig. 2.15. Details of joint sealing groove

2.8.5 Joint sealing

The purpose of sealing joints is to prevent dust, water and other debris entering the joint. The sealant must be able to withstand the strain of the opening of the joint resulting from contraction of the slab and remain fixed to the faces of the groove sides (Fig. 2.15). For floors which are to be trafficked by rigid-tyred vehicles and where joint widths are greater than 5 mm, it may be necessary to use a strong semi-rigid sealant such as pouring grade epoxy or grout in order to provide support to the edges of the joint. The low elasticity of such sealants might cause them to fail when applied to an active joint. For this reason they are applied at a later date when shrinkage movements have already occurred. Alternatively, more flexible mastic based materials can be used. In slabs where joint spacings exceed 10 m it is particularly important to carry out joint sealing. With deep crack inducing grooves a polyethylene backing rod may be placed in the lower part of the sealing groove to reduce the amount of sealant required (Fig. 2.15).

Durability can be improved with polysulphide sealants and when appearance is of concern, *gun-grade* materials can be used, especially for narrow grooves. Sealants should be inspected and maintained regularly.

2.9 Floor construction case study

2.9.1 Details of case study

Figure 2.16 shows a plan of an industrial building which is to include a single pour floor which will be subjected to patch loads from vehicles, uniformly distributed loads from materials storage and point loads from the supporting columns of a mezzanine floor. Detailed calculations for the floor are presented in Chapter 5. The building includes several rooms which are to be constructed directly off the floor. Chapter 5 shows that a 225 mm thick floor on 250 mm thickness of granular sub-base material is suitable. A C40 concrete reinforced with 20 kg/m^3 of steel fibre is to be used for the floor slab. Figure 2.16 shows the grid lines which align with structural column centres. The building upon which this example is based was constructed in the UK in 1994.

Figure 2.17 shows the proposed arrangement of joints. The joint spacings are determined in Chapter 5. A perimeter beam circumscribes the floor so as to facilitate the use of a laser screeding machine. The perimeter beam is 700 mm wide and is widened at external column locations. The perimeter beam is separated from the floor by an isolation joint which allows the floor and the perimeter beam to move independently of each other. Each of the three internal columns is surrounded by a diamond-shaped pad which is separated from the floor by an isolation joint. The mezzanine floor supporting columns are to be bolted onto the floor and do not have independent foundations. The main structural columns and the building perimeter wall are to have independent foundations.

Figures 2.16 to 2.29 illustrate the steps involved in construction of such a structure.

2.9.2 Setting out

Figure 2.18 shows the grid lines which are established on the site prior to any construction work. Pegs should be located in the ground using accurate surveying instruments at each end of each grid line. Temporary bench marks should also be established. At this stage, it may become evident that the proposed levels of the building are inappropriate. For example, it may be realized that the proposed levels will lead to excessive cut or fill. It is often the case that a revision in levels is possible at this stage, although care needs to be exercised in relation to drainage falls and highway gradients. Note that at this stage no attempt is made to locate the position of the floor.

2.9.3 Site strip

Figure 2.19 shows how the topsoil and underlying material is removed down to the level of the underside of the sub-base. Usually, this involves removing material approximately 1 m beyond the extreme grid lines. Care needs to be exercised in the protection of the underlying material from weather and construction traffic, especially as the exposed surface will be flat and therefore subject to ponding. Many types of soil in the UK will be significantly weakened if trafficked in a saturated condition and, once weakened, may never recover their undisturbed strength. The floor designer will have relied upon a soil strength in structural calculations and failure to ensure that the strength is maintained may lead to failure of the floor. Should the excavation become waterlogged, it may be necessary to suspend construction.

2.9.4 *Excavate and construct pad footings for steel columns*

Pad footings of plan dimensions 1.5 m × 1.5 m are to be constructed by excavating below the site strip level. If the soil remains stable with a vertical edge, no formwork is required and the pad footings can be constructed directly against the soil. Wet blinding concrete is first poured into each excavation, steel mesh and/or bars are then fixed and the concrete is poured. Sockets are left in the surface of the pad footings to allow the fixing of the steel columns. The upper level of the foundations has to be sufficiently deep to permit both the floor and the foundation to the building perimeter wall to be constructed. In this case, the foundations are 750 mm thick and their upper surface is 300 mm beneath the lower surface of the floor sub-base. After casting, a check should be made that each base aligns accurately with the grid. In this case, each base requires 1.68 m^3 of concrete so one concrete delivery of 5 m^3 will be sufficient for three bases.

2.9.5 *Install underground services*

Services need to be installed and taken through the subgrade, then vertically upwards to the level of the floor. Backfilling of all trenches must be undertaken diligently to avoid later settlement which would induce stress in the floor slab. All water and gas pipe runs should be pressure tested prior to any trench reinstatement. The position of all services should be recorded to minimize disruption in case of future repairs.

2.9.6 *Erect structural steelwork*

The structural steelwork and the cladding must be installed before the floor can be constructed. Usually, the structure comprises hot rolled mild steel portal and gable frames supporting cold rolled galvanized steel purlins and side rails. Wind bracing may be installed in one or more bays and the column base plates are fixed permanently to their pad footing foundation bases.

2.9.7 *Fix roof and side cladding*

It is common for industrial buildings to include corrugated steel sheets as the roof cladding and as cladding for the upper parts of the sides. The cladding may include an inner lining material to provide thermal insulation and roof lights and windows may be included. More sophisticated cladding systems may be installed where end usage is predominantly office accomodation or high-tech workshop facilities. The cladding must prevent sunshine and wind from interfering with the floor construction operation.

2.9.8 *Excavate and construct external wall foundations, then external walls*

Most industrial buildings include a height of masonry or similar durable cladding material as the lower part of the building walls. This is partly to improve durability and partly to improve appearance. Two leaves are usually provided, an inner leaf of concrete blocks constructed in discrete bays between the flanges of the columns (sometimes including a box around the columns) and an external leaf of brickwork sailing past the columns externally to create an uninterrupted effect. The masonry is constructed on its own foundation comprising a strip footing constructed above the column pad footings. The concrete block leaf is 100 mm thick, the brickwork leaf is 105 mm thick and the cavity between them is 50 mm thick, so the total wall thickness is 255 mm. The inner face may be plastered or dry lined with plaster board, adding a further 10 mm thickness.

2.9.9 *Construct sub-base*

The crushed rock or cement stabilized sub-base is now installed by spreading the material and compacting it with a vibrating roller. The material is installed right up to the masonry walls and a plate vibrator is needed to compact adjacent to the walls to avoid damage. The plate vibrator is also required around service ducts. Compacted crushed rock materials have a variety of surface textures. In the case of harder or coarser grained materials, the surface may be open textured or 'hungry', in which case it should be blinded with limestone dust to create a closed surface. In the case of cement stabilized sub-bases, curing will be required comprising an impermeable spray. Although the sub-base surface can be used by construction traffic, care must be taken to avoid rutting the surface of the material. This can occur in the case of some softer limestone crushed rock materials, for example dolomite won from some Durham quarries.

2.9.10 Construct perimeter strip and isolation pads around internal columns

The perimeter strip is approximately 700 mm wide and the same depth as the floor — 225 mm in this case. It is separated from the masonry walls and the columns by a full movement joint. It can be constructed from the same fibre reinforced concrete as the main floor or can be reinforced with longitudinal bars. Joints should be provided according to the reinforcement percentage as for the main floor. The perimeter strip joints need not align with the main floor joints, indeed, they should not be at column positions, otherwise points of weakness will occur. Figure 2.26 shows diamond surrounds to the columns. Alternatively, circular surrounds may be provided. A full movement joint should be provided around each isolation pad and around the inside of the perimeter strip. This joint should comprise a 12 mm wide piece of compressible material fixed the full depth of the perimeter strip. A sacrificial timber or synthetic strip is installed along the upper 15 mm of the perimeter beam, to be removed later and replaced with a permanent sealer.

2.9.11 Install concrete floor slab

If a slip membrane has been specified, it should be installed first, care being taken to avoid folds or ripples which might weaken the concrete slab. The concrete is then spread, compacted and finished by laser screed, with powertrowelling taking place after the concrete has set but before it has gained strength. Often, powertrowelling takes place between 8 h and 16 h after the concrete is mixed. Fibres can be added to the concrete at the mixing plant or alternatively at the site by emptying fibre bags into the readymix truck and mixing for two minutes. The readymix trucks usually discharge at the intended position where the concrete will form the floor. If access is not possible, the concrete can be pumped up to a distance of 100 m.

2.9.12 Cut 40 mm deep grooves to induce joints

Grooves must be cut as soon as the concrete will allow the operation to take place without forming a ragged edge to the cuts. Usually cutting can commence approximately 24 h after mixing the concrete. Cutting should be completed within 48 h of mixing in order to avoid the risk of uncontrolled cracking. A 3 mm wide, vertically sided groove is formed using a grinding wheel. The joints are cleaned out using a high pressure air line and a sealer with an expansion capability of 30% or more is applied to the joints, including the full movement joints at the perimeter and around the internal columns – the sealer replaces the sacrificial strip placed above the compressible material.

2.9.13 Construct internal walls and mezzanine floor support columns

Because the internal walls and mezzanine floor support columns are constructed onto the surface of the floor, they are left until last. The walls are constructed conventionally, with a layer of polyethylene being placed between the floor and the underside of the wall to act as a moisture barrier. The mezzanine floor columns are bolted to the floor using expanding bolts. Shims may be needed to ensure that the columns are vertical.

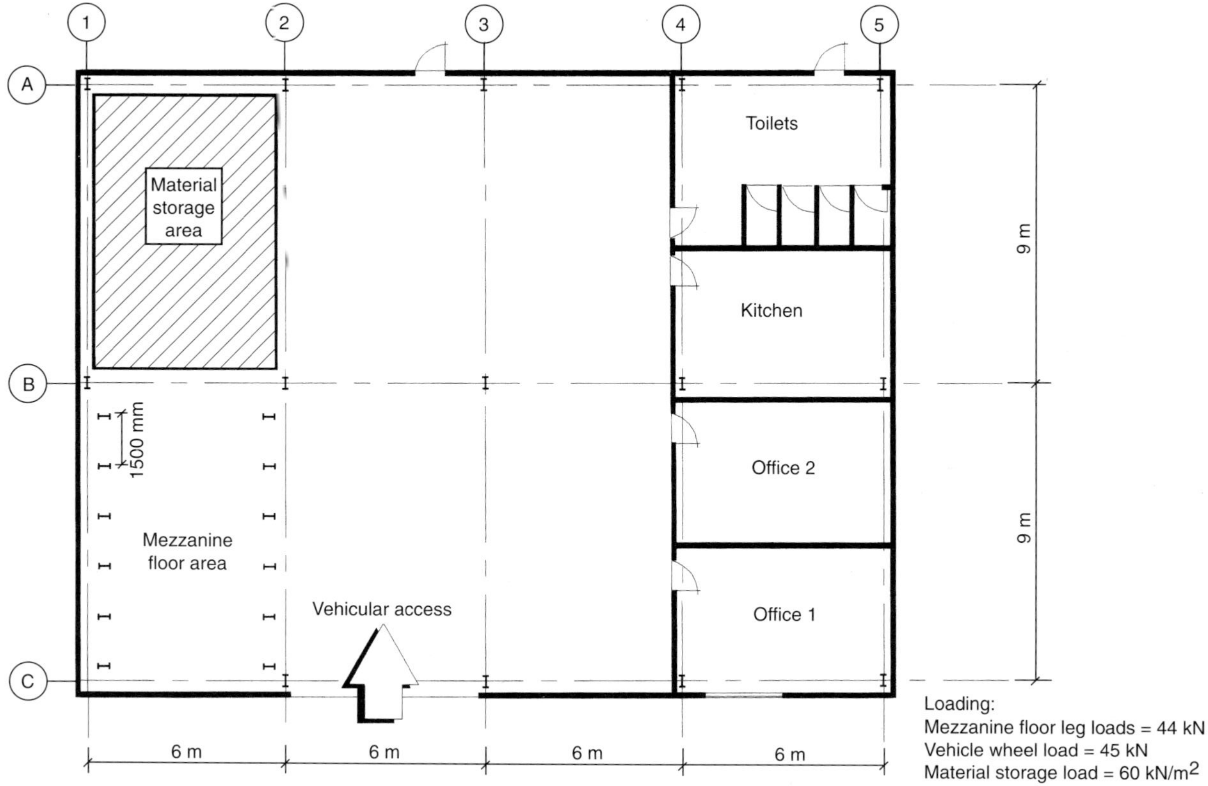

Fig. 2.16. Industrial unit design example

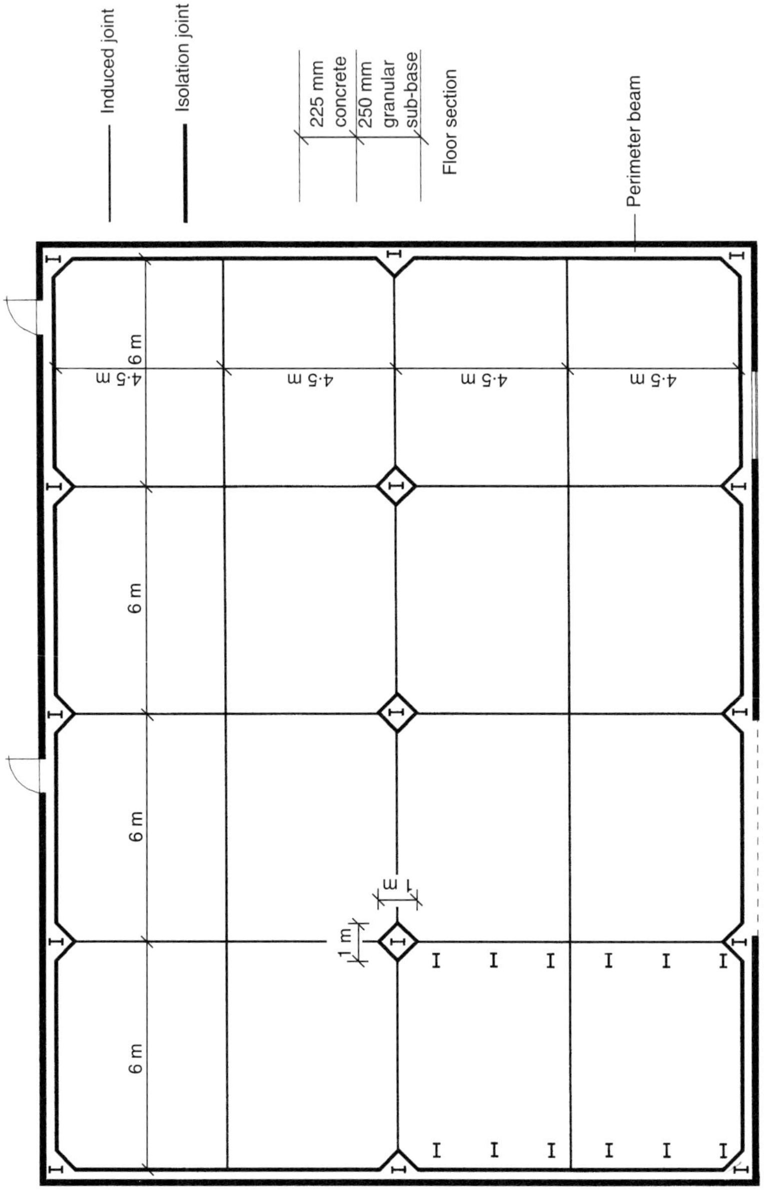

Fig. 2.17. Arrangement of joints in design example

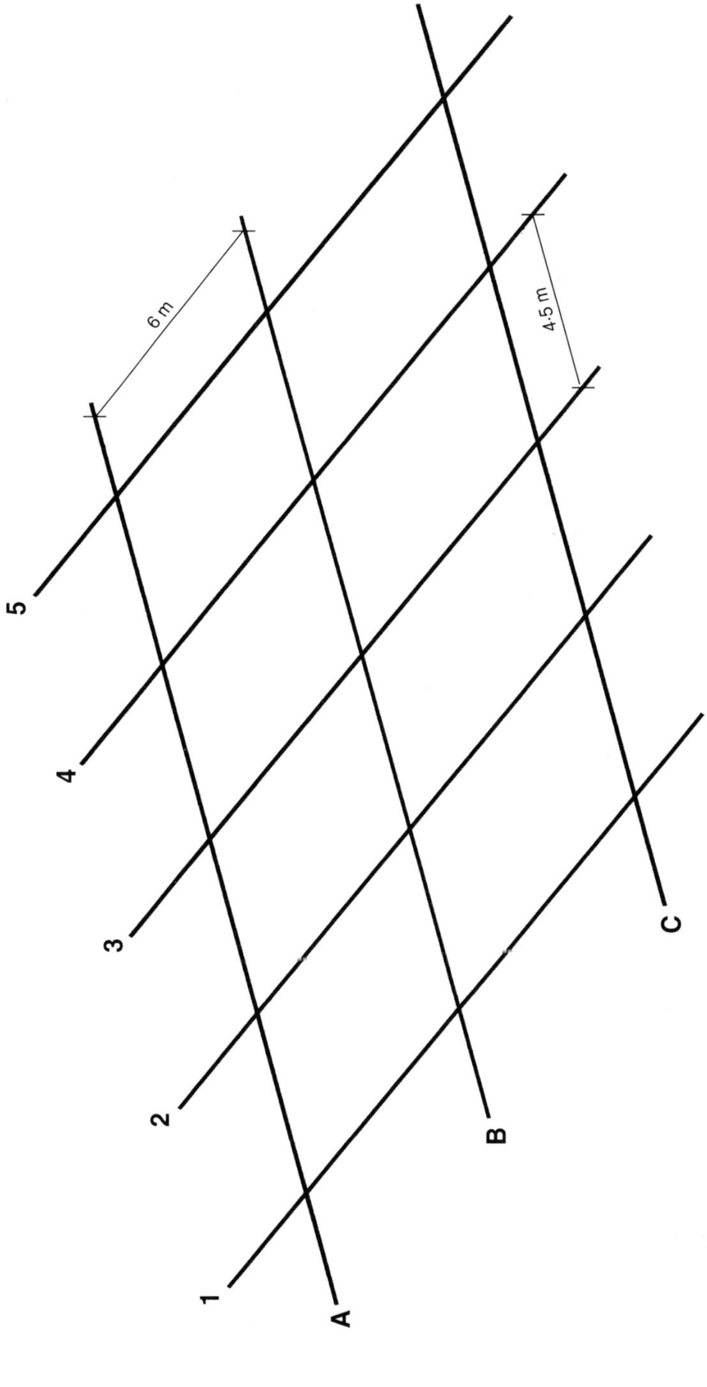

Fig. 2.18. Establish grid lines on site

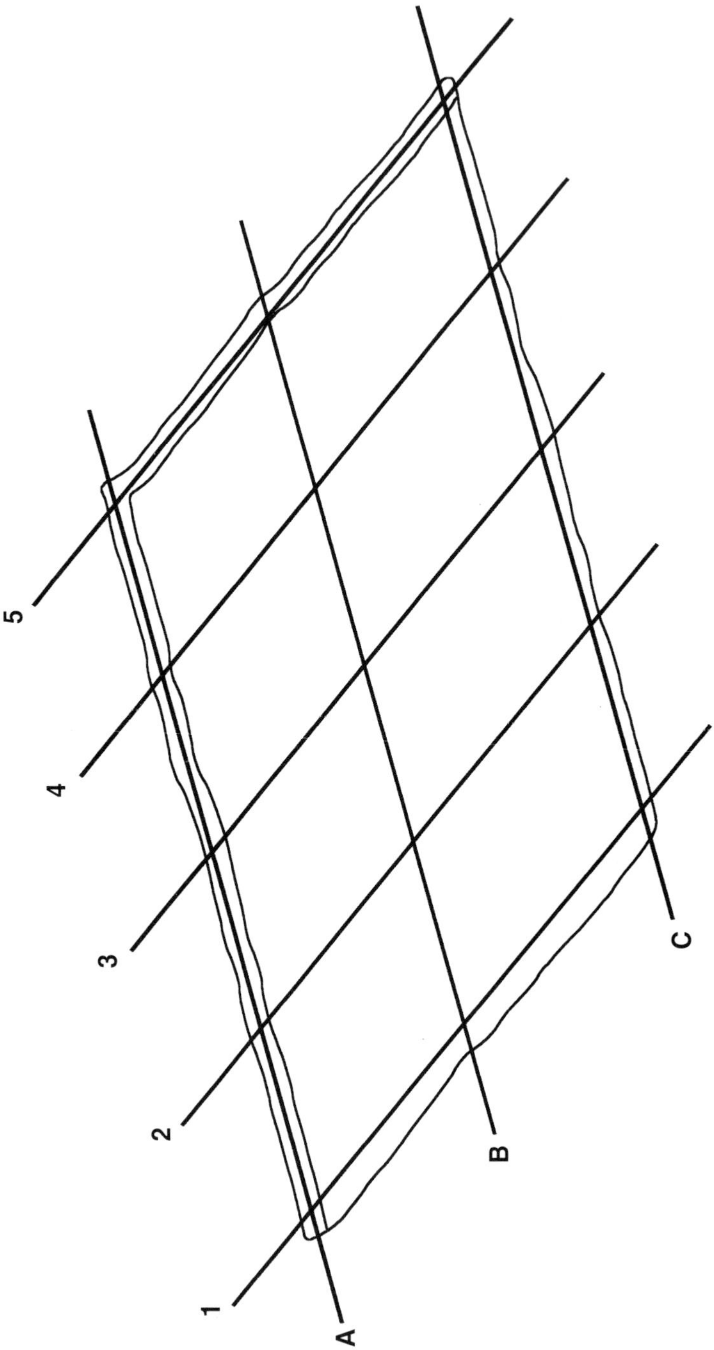

Fig. 2.19. Topsoil strip — excavate to level of underside of sub-base

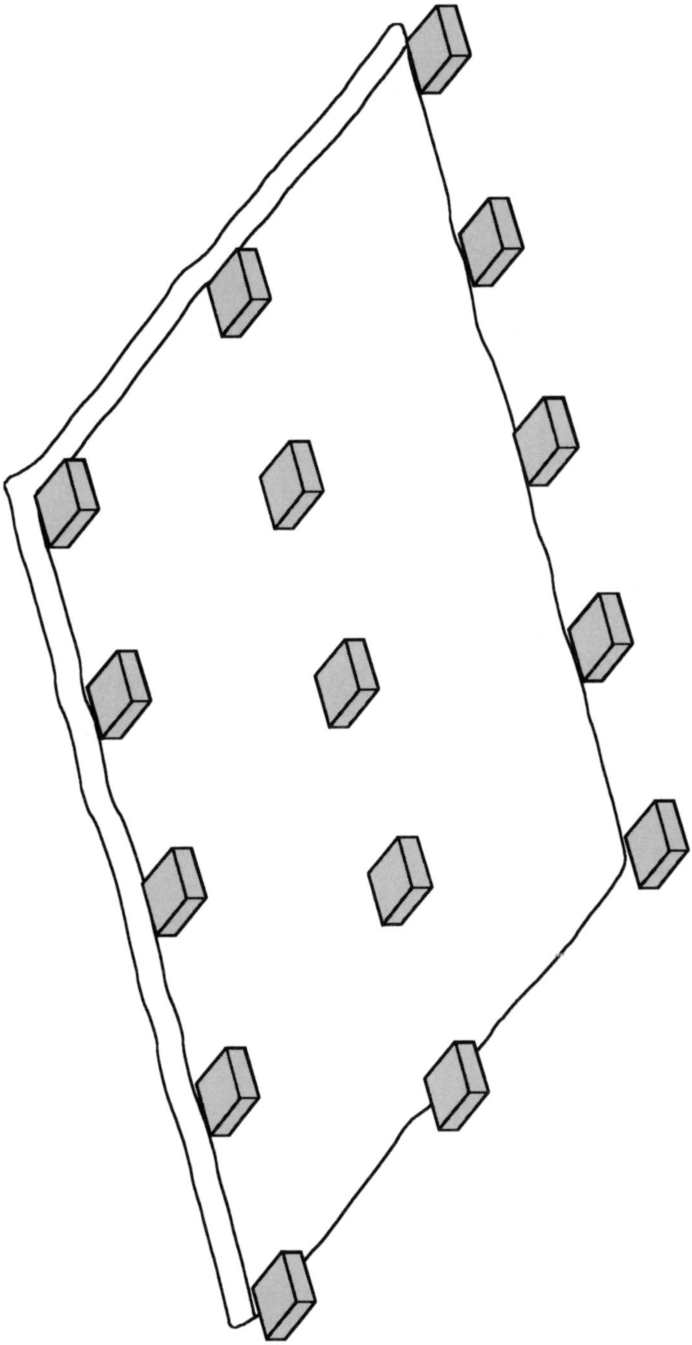

Fig. 2.20. Excavate and construct pad footings for steel columns

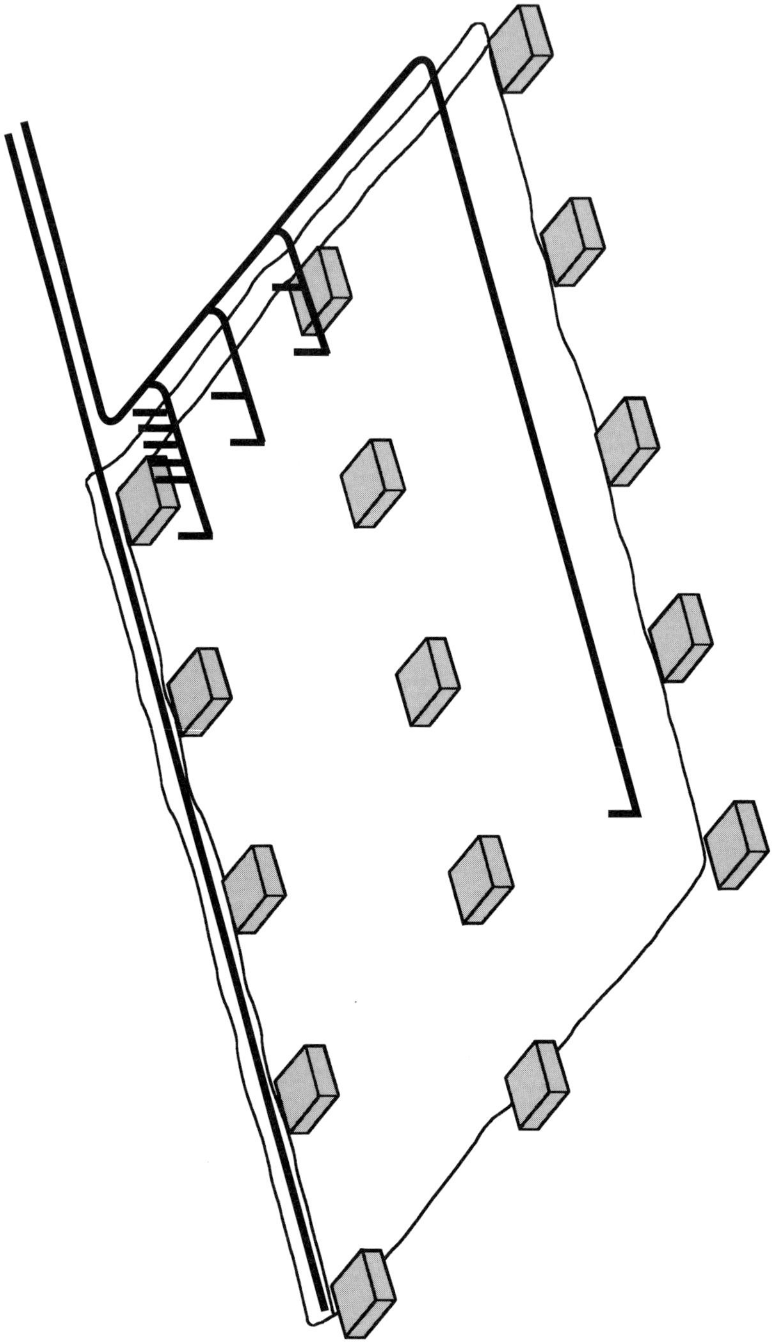

Fig. 2.21. Install ducts and pipes for gas, electricity, water, foul drainage, surface water runoff and communications

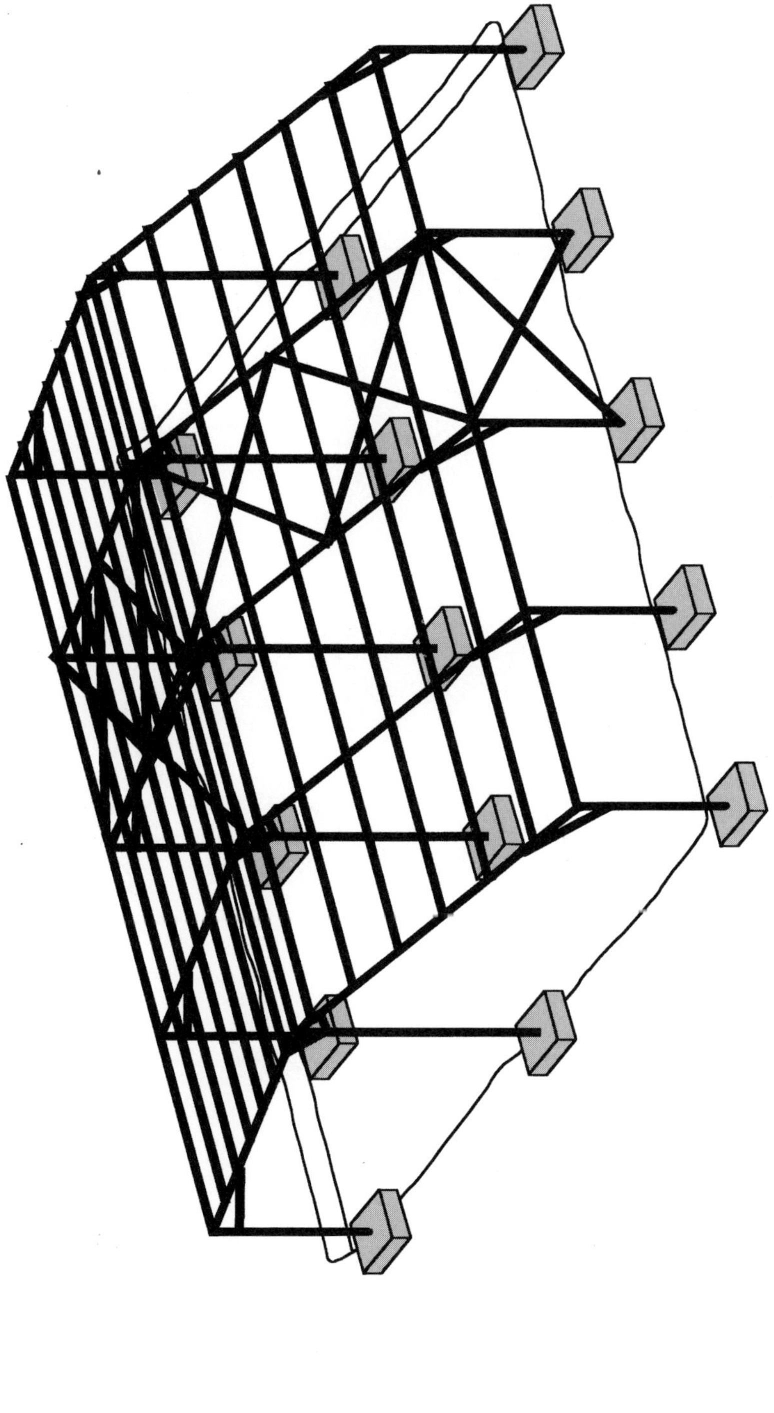

Fig. 2.22. *Erect structural steelwork*

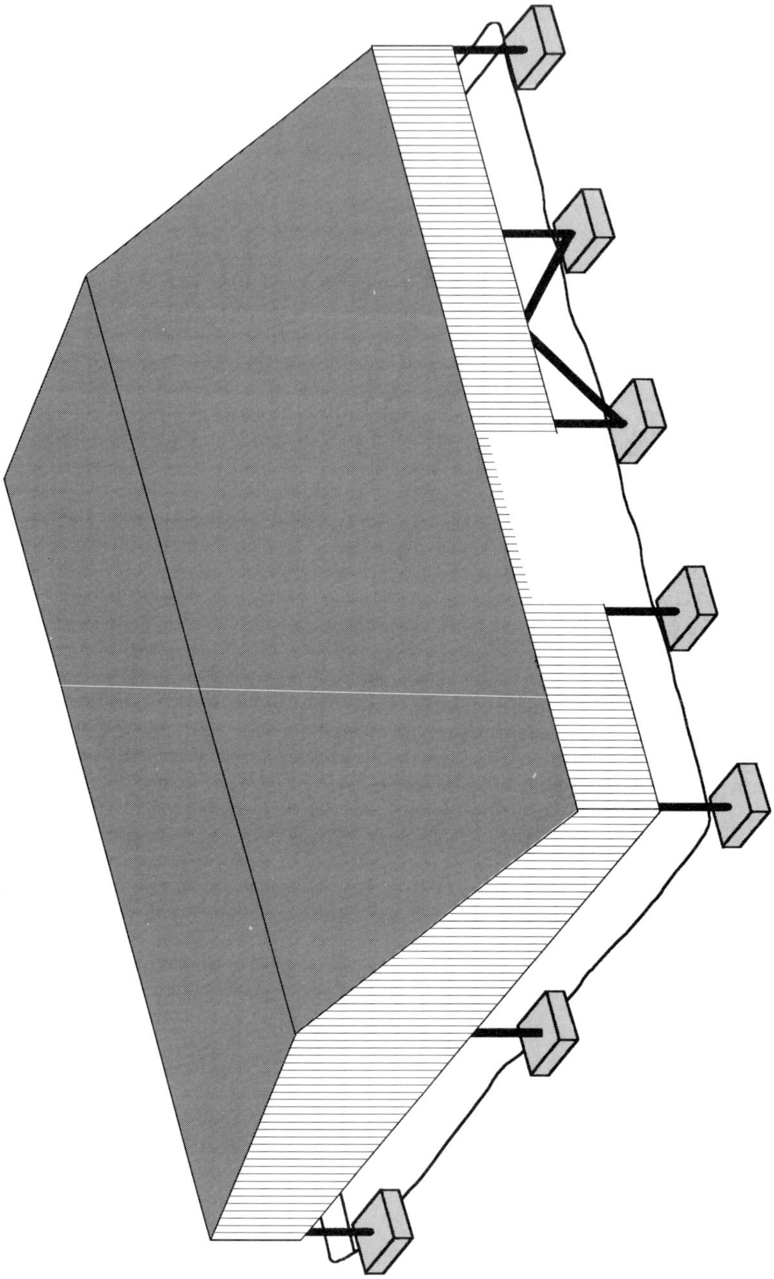

Fig. 2.23. Fix roof and side cladding

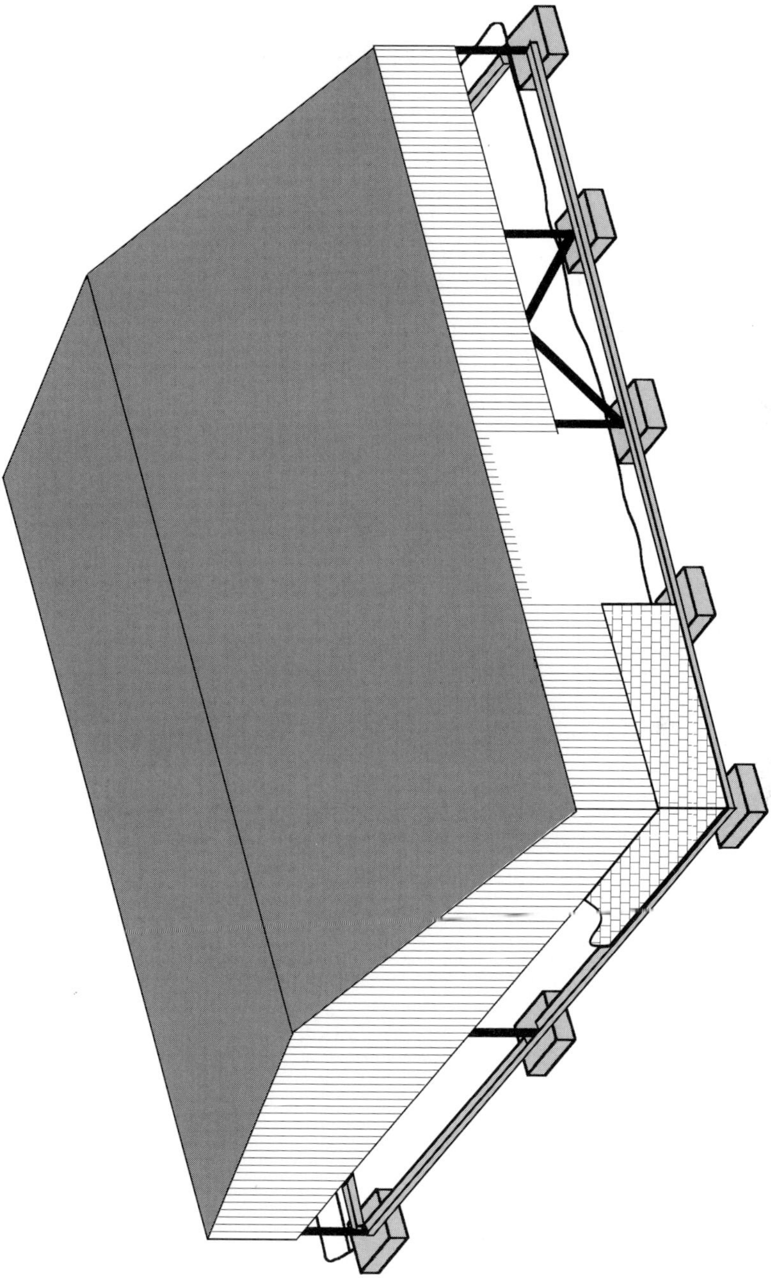

Fig. 2.24. Excavate and construct external wall foundations, then external walls

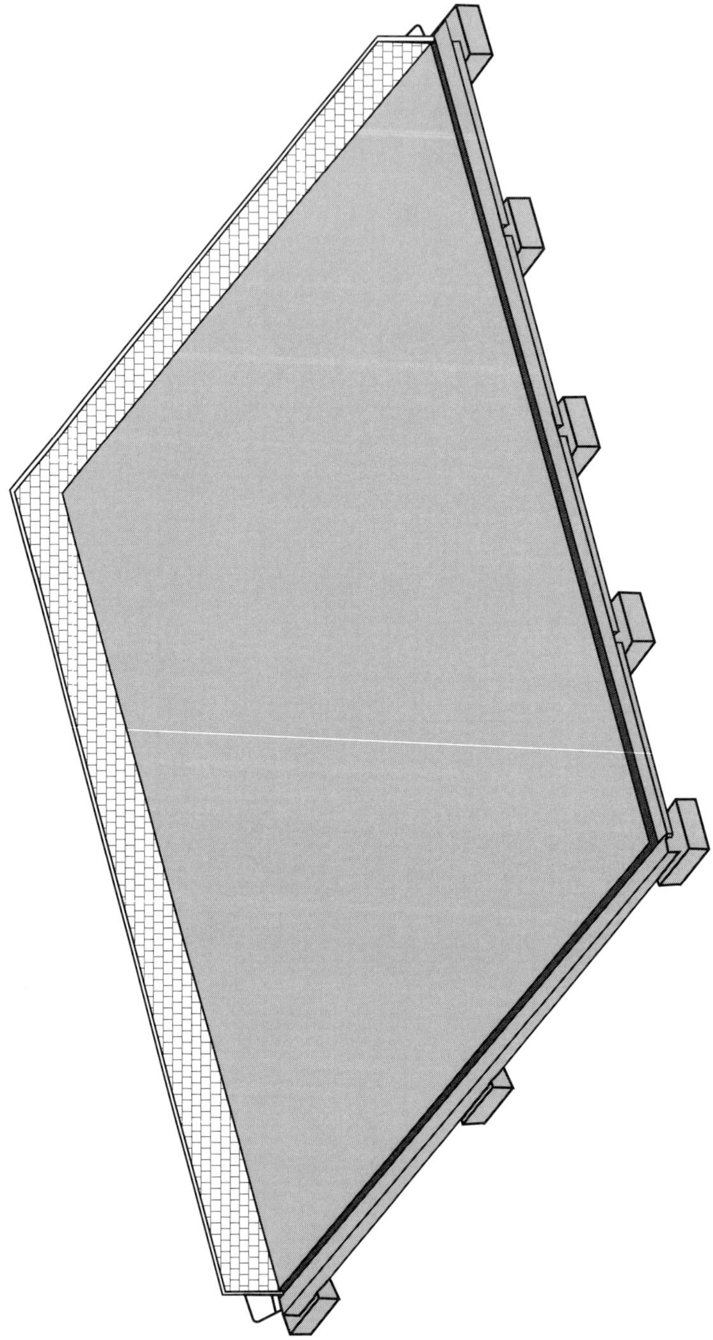

Fig. 2.25. Construct sub-base to floor (structure, cladding and nearside walls not drawn for clarity)

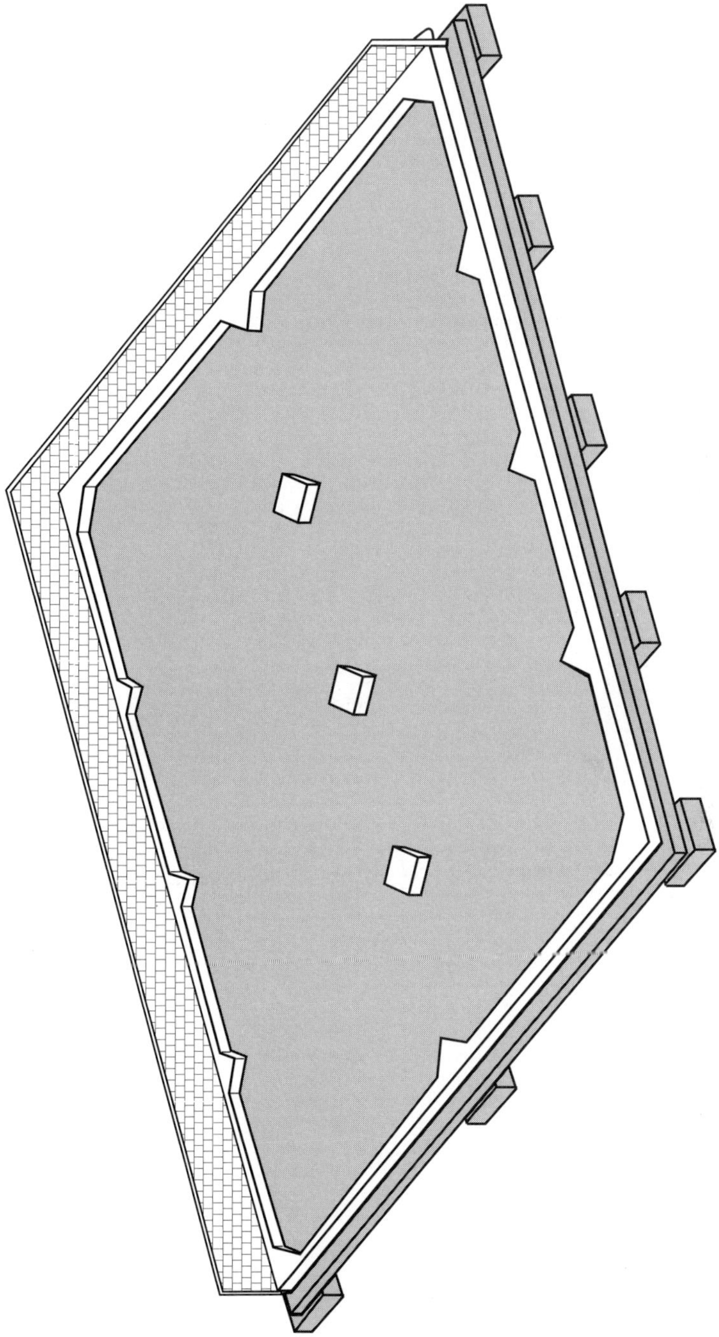

Fig. 2.26. Construct perimeter strip and isolation pads around internal columns (nearside walls omitted for clarity)

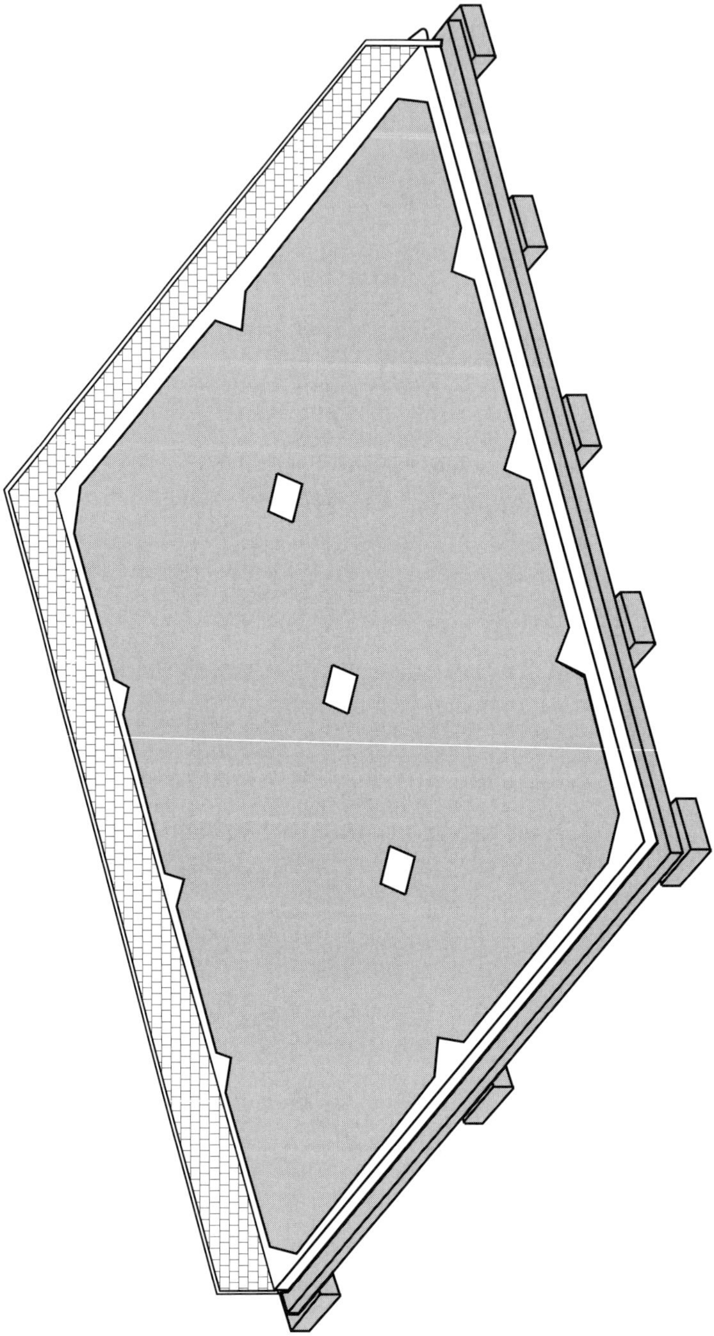

Fig. 2.27. Install floor slab

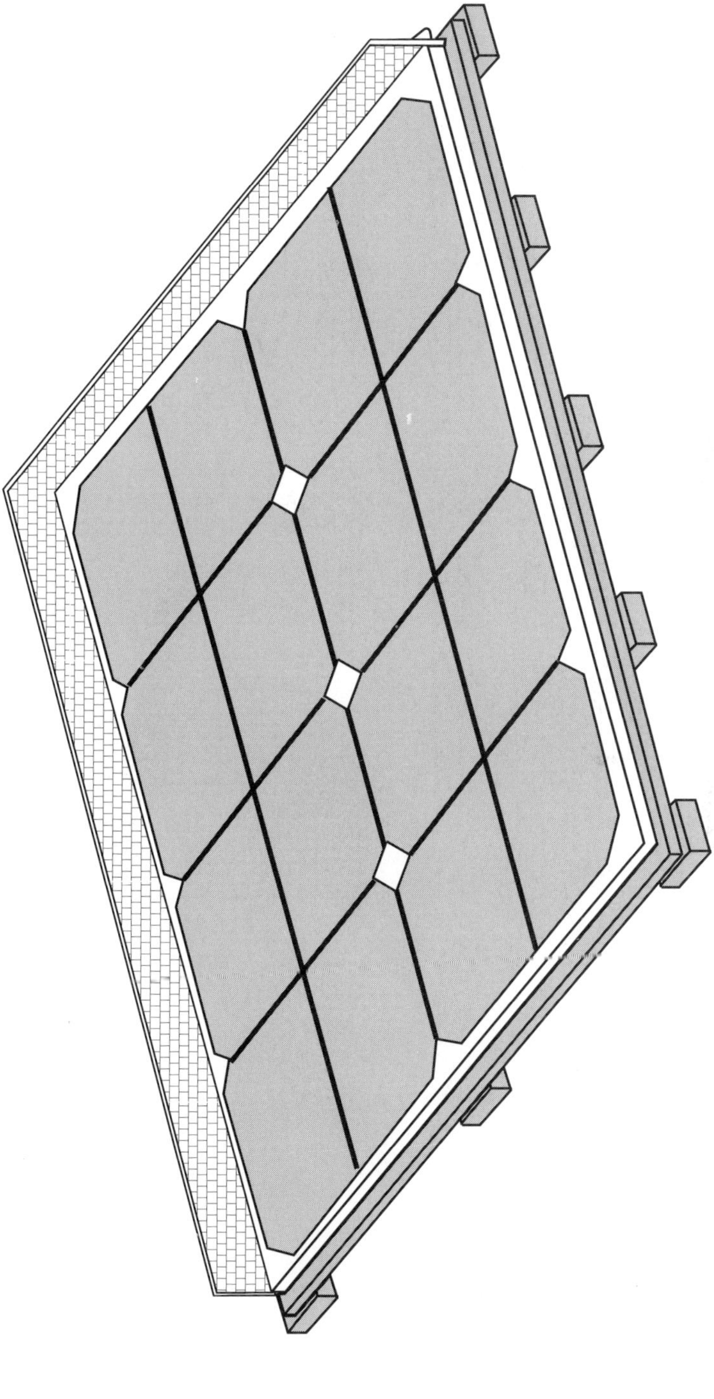

Fig. 2.28. Cut 40 mm deep grooves to induce joints

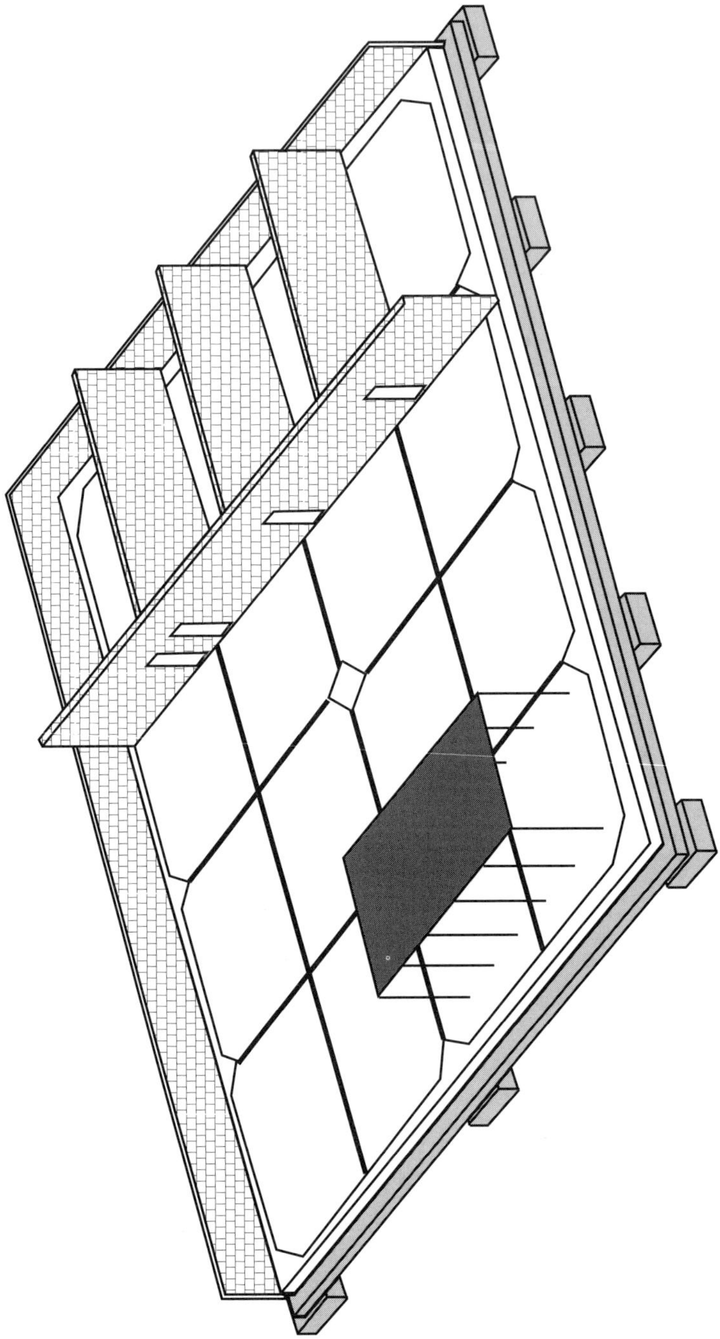

Fig. 2.29. Construct internal walls and mezzanine floor

3. Loading

3.1 Loading considerations

Loading of industrial floors will be one of or a combination of the following:

- Uniform Distributed Load (UDL)
- Live load (LL)
- Point Load (PL)
- Contact Pressure (CP)
- Horizontal Load (HL).

Floor design comprises assessing the loading which the floor is predicted to sustain and selecting materials, thicknesses and a joint configuration which will sustain those loads whilst at the same time satisfying flatness, durability, abrasion and riding quality requirements. An essential part of the design process is the assessment of the anticipated load regime. Frequently the end use of the floor will be uncertain at the time of floor design and construction. In such cases it is common for a distributed load and a point load to be specified and for all of the floor to be capable of withstanding the selected values. Horizontal load may be introduced into the floor if the building frame relies upon the floor as a ground level tie or if vehicles undertake turning manoeuvres. In such cases, special care is needed when detailing joints which must be able to transmit tension between neighbouring bays. In the case of tied portals, the value of the horizontal load can be calculated by undertaking an analysis of the building frame, taking the horizontal reaction at the foot of each structural column and distributing the load into the slab. It may be necessary to detail reinforcing bars radiating from column feet to ensure that the true behaviour of the horizontal forces approximates the design assumptions closely. A common category of loading comprises leg loads applied by storage systems or mezzanine floor columns. Different types of storage systems are considered explicitly. Often, floors used for storage require particularly tight tolerances.

3.2 Unit pallet racking

For warehouse storage, the standard Euro Pallet (0.8 m × 1.0 m) has a loading range of 5 kN to 20 kN with a storage height of between 1.2 m and 1.4 m (see Fig. 3.1).

Some multi-national storage companies have designed alternative pallets to the standard Euro Pallet. The modified pallet tends to be longer for ease of storing larger products (Fig. 3.2). The only dimensional change to the modified pallet is its length since all pallets have to be transported on the same vehicles for economic reasons.

Fig. 3.1. The standard Euro Pallet

Fig. 3.2. A modified pallet and a Euro Pallet

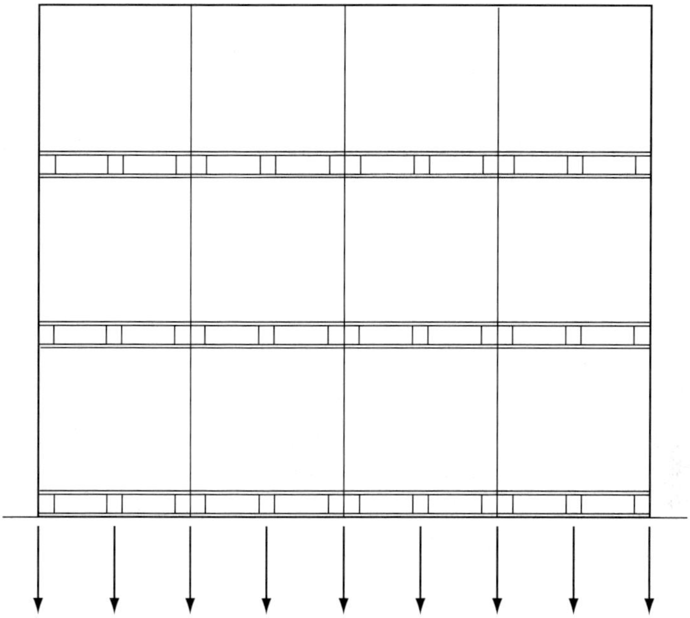

Fig. 3.3. Typical block stacking

3.3 Block stacking

Block stacking (Fig. 3.3) is usually limited to the width of aisles, size of the fork lift and the actual weight of the pallet plus load. The stacking process is based upon placing pallet loads directly on top of each other (materials ranging from paper to steel and where handling is not of great importance). The maximum height is around 9.5 m. Settlement can take place to a greater extent than with other loading regimes. Uplift can be a major problem and can be seen especially if there are mid-aisle joints: hence the monitoring of unit pallet load stacking is recommended.

3.4 Pallet storage racking

Racking (Fig. 3.4) allows storage to a greater height and facilitates access to intermediate pallets. Adjustable pallet racking is the usual design used and comprises a steel framework where the pallets rest on steel beams. The racks are often placed back-to-back for maximum access for the fork lifts. The height of the warehouse always dictates the number of levels of racking with adequate operating clearance for each individual pallet. The aisle widths vary from 2.7 m (Fig. 3.5) for free ranging fork lift trucks to 1.2 m (Fig. 3.6) for narrow aisle trucks. Figure 3.7 illustrates a racking fixing foot.

Fig. 3.4. Typical pallet storage racking

Fig. 3.5. 2.7 m wide storage racking

Fig. 3.6. 1.2 m wide storage racking system. There are two fixing feet per bay (except at the end) where the whole load is transmitted to the floor

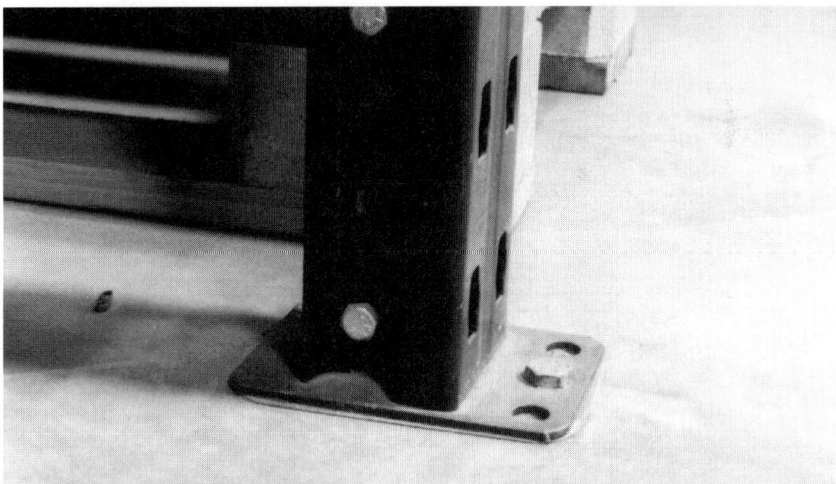

Fig. 3.7. Racking fixing foot. If the racking is high the loading on each leg can be substantial. The load may exceed 300 kN but the critical stress beneath the legs has no bearing on the width of the aisle. To prevent damage to concrete floors and to help distribute the loading, shims and footplates are often specified with racking systems

3.5 Pantograph racking

Pantograph racking (Fig. 3.8) is similar to standard storage racking but with extra horizontal depth. Pantograph racking is two pallets deep giving a greater density of storage and a greater concentration of point loads.

3.6 Raised storage platforms (mezzanine floors)

Mezzanine floors are classed as 'free-standing steel floors' which give extra storage and optimize the use of headroom above ground level. Construction is often in a grid formation (approximately 4 m × 3 m) supported by steel posts each having typical leg loads of between 30 kN and 75 kN. Loads greater than 75 kN are usually taken down to a foundation beneath the floor.

3.7 Mobile pallet racking

Mobile pallet racking comprises rows of racking mounted on rails which are driven by an electric motor. Usually limited to 6 m high, the system is very efficient in the use of floor space (approximately 80%) and every pallet can be accessed by moving the racks along the rails. Being mounted

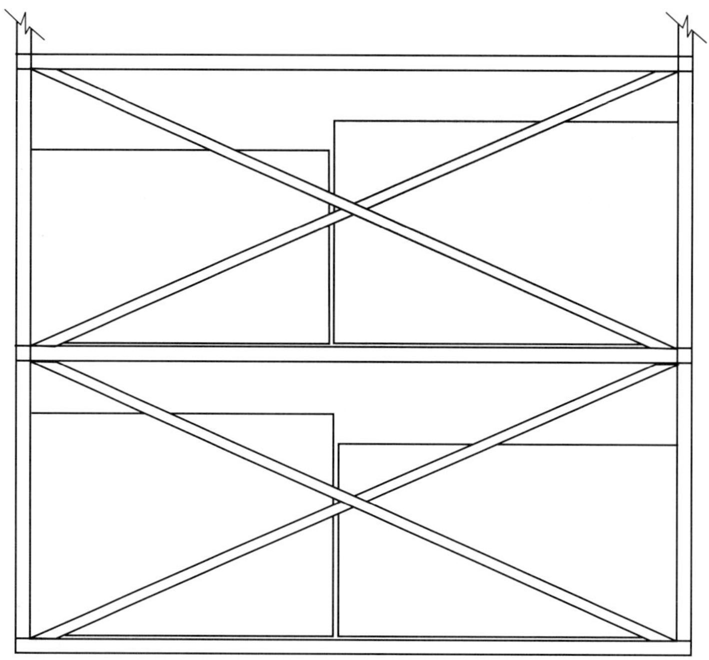

Fig. 3.8. Pantograph racking

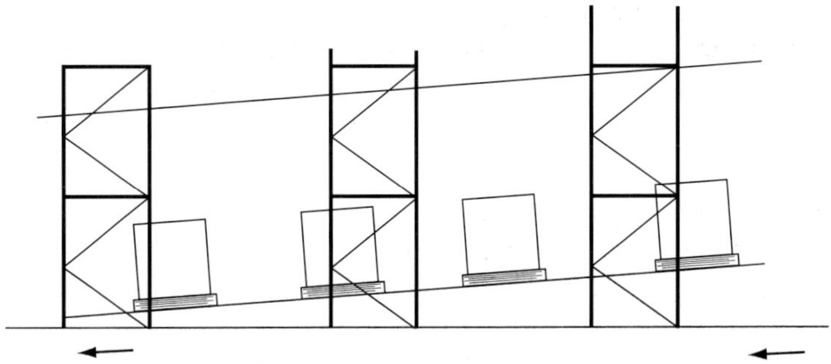

Fig. 3.9. Live storage racking

on rails, this racking system often applies large line loads of the magnitude of 23 kN/m.

3.8 Live storage systems

Live storage racking (Fig. 3.9) comprises a last-in first-out rotation storage system which operates by placing a pallet at one end of the racking which is on a downward conveyor. Gravity moves the pallets to the opposite end of the system to the output location where distribution takes place.

3.9 Drive in racking

Each pallet is supported on cantilever brackets which are bolted to the racking steelwork. This system has no permanent access aisles. Depending on the closeness of the legs, with a 20 kN pallet the maximum load on the racking is 40 kN leg loads for five storeys high.

3.10 Shelving

Shelving systems store smaller products (Fig. 3.10). The shelves are often of uniform cross-section and are placed back-to-back which results in concentrated leg loads.

3.11 Mobile handling equipment

Mobile transporters are required with many storage systems and they impose loads through the contact areas of their tyres. Pallet transporters (Fig. 3.11) comprise hand-pushed trailers carrying a maximum load of 20 kN on very small wheels with an average contact pressure of 9 N/mm^2.

Counterbalance fork lift trucks (Fig. 3.12) are front loading telescopic mast vehicles with an average load of 30 kN. Height is usually limited to

8.5 m for stability reasons and the maximum wheel load occurs from the rear wheels when the truck is unladen.

Reach trucks (Fig. 3.13) have a moving telescopic mast or pantographic load extender and often operate in very narrow aisles with a maximum load of 30 kN and height limited to 8.5 m.

Fig. 3.10. Typical shelving system

Fig. 3.11. Pallet transporter

Fig. 3.12. Counterbalance fork lift truck

Fig. 3.13. Reach truck

Within very narrow aisles rubber stoppers line the aisle to stop the transporter hitting and damaging the products already in storage. The sideloader reach truck has a telescopic mast at right angles to the driver's cabin. Very narrow aisle trucks can transport pallet loads in any direction because of their rotating traversing mast. These trucks can be wire or rail guided but usually they move freely. Within a 1.3 m wide aisle a load of 20 kN can be lifted up to a maximum 12 m. Order pickers carry only empty pallets to a height of 9.5 m where the operator can stack by hand. Stacker cranes run on rails bolted to the floor slab. This system gives a possible lifting height of 30 m.

3.12 Methods of specifying floor loading

3.12.1. Contact pressure

Note that rack footplates, as shown in Fig. 3.14, are designed to accommodate the fixing bolts rather than to distribute the load so the nominal floor contact area is relatively small.

3.12.2. Average contact pressure

From TR 34,[3.1] contact pressure values (in N/mm^2) for typical storage systems and equipment are:

- 30 kN Counterbalance lift truck 2.4
- 20 kN Reach truck 5.6
- 20 kN Pallet truck 11.0

Fig. 3.14. Typical footplate with fixing bolt

Table 3.1. Load classification

Load classification	Typical load: kN	Usual type of loading
Light	42	Pallet racking Mezzanine floor Shelving Fork lift
Medium	60	Pallet racking Mezzanine floor Shelving Fork lift
Heavy	87	Pallet racking Mezzanine floor
Very heavy	114	Pallet racking Mezzanine floor

3.12.3. Horizontal loads

Horizontal loads applied by the braking of mobile materials handling equipment are usually ignored. When the building frame uses the floor as a tie, it may be necessary to consider the tie force, both in the structural design of the floor slab and in the detailing of joints. In large floors, it may be that tie forces will be dissipated into the sub-base near to the perimeter of the floor. The transfer of horizontal tie forces into the sub-base can be assessed by considering the vertical force applied by the self-weight of the floor slab and applying a coefficient of friction according to whether a slip membrane is present.

3.12.4. Loading table

In the design of concrete formless floors loading is a major consideration and Table 3.1 can be used to assess loads in situations where no further information is available.

4. Design

4.1 Introduction

The design procedure set out in this chapter comprises calculating stresses resulting from the loading regime and ground conditions and comparing those stresses with the strength of the concrete. Because floor design is an exercise in serviceability, the design procedure is based upon ultimate load analysis using partial safety factors for materials and load appropriate to serviceability. Concrete characteristic flexural strength is used as the basis of stress assessment and load induced stresses are computed using classical methods. The following factors have to be taken into account in ground floor design:

(a) loading regime;
(b) strength of concrete;
(c) strength of existing ground and effect of the sub-base.

These are now each considered in turn.

4.2 Loading regime

Loading has been described in Chapter 3. Fatigue often leads to floor distress and in some cases a fatigue factor is built into the design method by the load safety factor. A fatigue factor of 2.0 is recommended for an infinite number of load repetitions and may therefore be used conservatively in all cases. The position of the load relative to the slab edge is critical and three alternative cases may need to be considered, namely internal loading (greater than 0.5 m from edge slab), edge loading and corner loading. With internal or edge loading the maximum stress occurs beneath the heaviest load at the underside of the slab. Corner loading creates tensile stress at the upper surface of the slab a distance away from the corner. This distance can be calculated from:

$$d = 2[(2^{0.5})rl]^{0.5} \qquad (4.1)$$

where r is the radius of the loaded area (in mm), l is the radius of relative stiffness (in mm) and d is the distance from the slab corner to the position of maximum tensile stress (in mm).

4.3 Strength of concrete

Formless floor slabs are frequently constructed from C30 or C40 concrete with a minimum cement content of 300 kg/m^3 with a slump of 50 mm or less. Design is based upon comparing concrete characteristic flexural strength with calculated flexural stresses whereas specification is by characteristic compressive strength. Table 4.1 shows flexural strength values for a range of commonly used concretes. Because floors fail by becoming unserviceable, a partial safety factor of 1 can be applied to characteristic strength in floor design.

4.4 Strength of existing ground and effect on sub-base

The design method requires a value for the modulus of subgrade reaction (K) which defines the deformability of the material beneath the floor. The following four values of K are used in the design procedure:

- $K = 0.013$ N/mm^3, very poor ground
- $K = 0.027$ N/mm^3, poor ground
- $K = 0.054$ N/mm^3, good ground
- $K = 0.082$ N/mm^3, very good ground (no sub-base needed).

Table 4.1. Mean and characteristic 28 day flexural strength values for various concrete mixes

	Flexural strength: N/mm^2	
	Mean	Characteristic
Plain C30 concrete	2.0	1.4
Micro-silica C30 concrete	2.4	1.7
C30 concrete 20 kg/m^3 ZC 60/1.00 steel fibre[a]	2.8	2.0
C30 concrete 30 kg/m^3 ZC 60/1.00 steel fibre	3.2	2.2
C30 concrete 40 kg/m^3 ZC 60/1.00 steel fibre	3.8	2.7
Plain C40 Concrete	2.4	1.7
Micro-silica C40 Concrete	2.8	2.0
C40 concrete 20 kg/m^3 ZC 60/1.00 steel fibre	3.2	2.2
C40 concrete 30 kg/m^3 ZC 60/1.00 steel fibre	3.6	2.5
C40 concrete 40 kg/m^3 ZC 60/1.00 steel fibre	4.2	3.2

[a] ZC 60/1.00 refers to a commonly used anchored bright wire fibre of total length 60 mm and wire diameter 1.00 mm

The beneficial effect of a granular sub-base is taken into account by increasing K according to the thickness and strength of the sub-base as shown in Table 1.5, Section 1.2.3.

4.5 Stress in concrete

The stress in a floor slab depends upon:

- the properties of the subgrade;
- the loading regime
  - uniformly distributed load (UDL);
  - patch/point loads;
- the thickness of the floor slab;
- the strength of the sub-base.

4.6 Design method for uniformly distributed loading

The usual loading system comprises alternate unloaded aisles and loaded storage areas. The maximum negative bending moment (hogging) occurs within the centre of the aisles and is given by:

$$M_{\text{hog}} = -\frac{q}{2\lambda^2}(B_{\lambda a'} - B_{\lambda b'}) \tag{4.2}$$

where a is the width of the aisle, b is the width of the loaded area, $a' = a/2$ and $b' = a/2 + b$.

The maximum positive bending moment (sagging) occurs beneath the centre of the loaded area (loading from adjacent blocks is ignored):

$$M_{\text{sag}} - q/2\lambda^2 B_{(1/2)\lambda b} \tag{4.3}$$

where q = UDL (characteristic load × load factor) (in N/mm^2), $B_x - e^{-x} \sin x$, $\lambda = (3K/Eh)^{1/4}$, K is the modulus of subgrade reaction, E is the concrete modulus (10 000 N/mm^2 for sustained load) and h is the slab thickness.

The two moment equations can be simplified into a single conservative equation for any combination of aisle width and stacking zone width:

$$M_{\text{max}} = -0.168q/\lambda^2 \tag{4.4}$$

The corresponding maximum flexural stress (in N/mm^2) is given by:

$$\sigma_{\text{max}} = 6M_{\text{max}}/h^2 \tag{4.5}$$

$$= 1008q/(\lambda^2 h^2) \tag{4.6}$$

where the maximum flexural strength cannot exceed the relevant characteristic value from Table 4.1. To ease calculation, Table 4.2 shows values of $\lambda^2 h^2$ for common combinations of modulus of subgrade reaction K and slab thickness h.

4.6.1. *Uniformly distributed load example*

Consider a 150 mm thick floor slab carrying a uniformly distributed load of 50 kN/m² between aisles on poor ground. As stated above, the maximum flexural strength (in N/mm²) is given by

$$\sigma_{\max} = \frac{1.008q}{\lambda^2 h^2}$$

and UDL (in N/mm²) is

$$q = 0.05 \text{ N/mm}^2 \times 2 = 0.10$$

In the above equation, 2 is the fatigue factor. A value of 2 means a load of 0.05 N/mm² can be repeated indefinitely.

For poor ground $K = 0.027$ N/mm³.

From Table 4.2:

$$\sigma_{\max} = 1.008\left(\frac{0.10}{0.035}\right) = 2.88 \text{ N/mm}^2 \quad (4.7)$$

From the characteristic strength values in Table 4.1, this stress can be withstood by a C40 concrete (incorporating 40 kg/m² of 60 mm long 1 mm diameter anchored steel fibre which can be used for any combination of aisle and stacking zone width.

Table 4.2. Values of $\lambda^2 h^2$ for combinations of slab thickness and modulus of subgrade reaction

Modulus of subgrade reaction K: N/mm³	$\lambda^2 h^2$ for slab thickness of (in mm):				
	150	175	200	225	250
0.082	0.061	0.066	0.070	0.074	0.078
0.054	0.049	0.053	0.057	0.060	0.063
0.027	0.035	0.038	0.040	0.043	0.045
0.013	0.024	0.026	0.028	0.029	0.031

4.7 Design method for patch loads using Westergaard equations

Highway vehicles and handling equipment apply loads to the surface of a floor as *patch loads*. In most cases, sufficient accuracy is gained by assuming the patch to be circular and to apply uniform stress throughout the patch. These assumptions lead to minor errors — the contact patch shape in the case of a commercial vehicle is nearer to a rectangle than to a circle but to assume accurately shaped patch loads would preclude the use of Westergaard equations. The error in assuming constant stress circular loading is very small and is conservative (i.e. true analysis would lead to a slightly thinner floor). The maximum flexural tensile stress occurs at the bottom (or top, in the case of corner loading) of the slab under the heaviest wheel load. The maximum stress under a patch load can be calculated by Westergaard[4.1] and Timoshenko equations as follows.

(a) Patch load in mid-slab (i.e. more than 0.5 m from slab edge):

$$\sigma_{\max} = \frac{0.275(1+\nu)}{h^2} P \log\left(\frac{0.36Eh^3}{Kb^4}\right) \tag{4.8}$$

(b) Patch load at edge of slab:

$$\sigma_{\max} = 0.529(1+0.54\nu)\frac{P}{h^2} \log\left(\frac{0.20Eh^3}{Kb^4}\right) \tag{4.9}$$

(c) Patch load at slab corner:

$$\sigma_{\max} = \frac{3P}{h^2}\left[1 - \left(\frac{1.41b}{\{Eh^3/[12(1-\nu^2)K]\}^{0.25}}\right)^{0.6}\right] \tag{4.10}$$

where

$\sigma_{\max}$ = flexural stress (in N/mm^2)
P = point load (in N) (i.e. characteristic wheel load × fatigue factor)
ν = Poisson's ratio, usually 0.15
h = slab thickness (in mm)
E = elastic modulus, usually 20 000 N/mm^2
K = modulus of subgrade reaction (in N/mm^3)
b = radius of tyre contact zone (in mm)
= $(P/\pi p)^{1/2}$
p = contact stress between wheel and floor (in N/mm^2).

Twin wheels bolted side-by-side are assumed to act as one wheel transmitting half of the axle load to the floor. In certain cases wheel loads at one end of an axle magnify the stress beneath wheels at the opposite end of the axle, distance S away (S is measured between load patch centres). To calculate the stress magnification, the characteristic length (or radius of relative stiffness, l) has to be found from Equation (4.11).

$$l = \left(\frac{Eh^3}{12(1-\nu^2)K}\right)^{0.25} \tag{4.11}$$

Once Equation (4.11) has been evaluated, the ratio S/l can be determined so that Fig. 4.1 can be used to find M_t/P, where M_t is the tangential moment. The stress under the heaviest wheel is to be increased to account for the other wheel. This is calculated by adding an appropriate stress:

$$\sigma_{add} = \frac{M_t}{P}\frac{6}{h^2}P_2 \tag{4.12}$$

where P is the greatest patch load and P_2 is the other patch load.

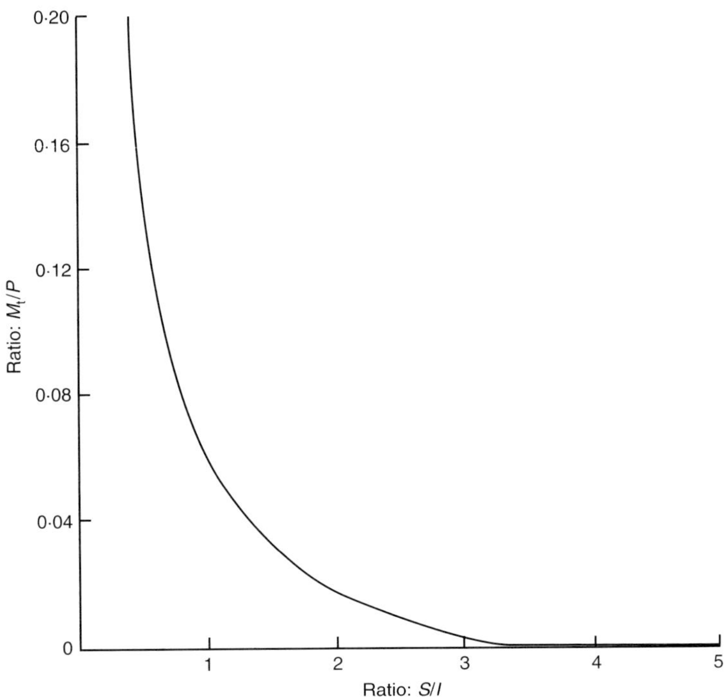

Fig. 4.1. Relationship between the ratio S/l and the ratio M_tP used in assessing the influence of load proximity

The stresses are then summed. It should then be verified that the flexural strength has not been exceeded (Table 4.1) for the prescribed concrete mix.

4.7.1 *Highway vehicle example using Westergaard equations*

Consider a highway vehicle with a rear axle load of 8000 kg and assume a slab thickness of 200 mm on good ground ($K = 0.054\ \text{N/mm}^3$). The fatigue factor is 2.0 so the floor can withstand an infinite number of such repetitions. The design axle load is 160 000 N (i.e. 80 000 N × fatigue factor (2.0)) so the design wheel load is 80 000 N. The following parameters are known or calculated:

p = contact stress between wheel and floor = $0.7\ \text{N/mm}^2$
b = radius of contact area = $(W/\pi p)^{1/2}$
= $(80\,000/\pi 0.7)^{1/2}$
=191 mm.

Substituting known values (E = elastic modulus = $20\,000\ \text{N/mm}^2$; μ = Poisson's ratio = 0.15) into Equation (4.9) yields:

$$\sigma_{\max} = 0.529(1 + 0.54 \times 0.15)\frac{80\,000}{200^2}\log\left(0.2\frac{20\,000 \times 200^3}{0.054 \times 191^4}\right)$$

$$\sigma_{\max} = 3.03\ \text{N/mm}^2 \quad \text{beneath one wheel.}$$

Assume the wheel at the other end of the axle to be 2.7 m away (S = 2700 mm). The radius of relative stiffness (l), using Equation (4.11) is:

$$l = \left(\frac{20\,000 \times 200^3}{12(1 - 0.15^2)0.054}\right)^{0.25} = 709\ \text{mm.}$$

Thus

$$\frac{S}{l} = \frac{2700}{709} = 3.8.$$

From Fig. 4.1

$$M_t/P = 0.005.$$

Therefore, using Equation (4.12) (repeated here for clarity)

$$\sigma_{\text{add}} = \frac{M_t}{P}\frac{6}{h^2}P_2$$

gives

$$= 0.005\left(\frac{6}{200^2}\right)80\,000$$

$$= 0.06\ \text{N/mm}^2.$$

The total stress is thus 3.03 + 0.06 N/mm^2 = 3.09 N/mm^2.

By comparing this stress with the characteristic strength of C40 concrete reinforced with 40 kg/m^3 steel fibres (3.2 N/mm^2), it can be seen that the proposed mix is satisfactory. A C40 concrete incorporating 40 kg/m^2 60 mm long 1 mm diameter anchored steel fibres would be inadequate for this design. However, the inclusion of a 250 mm thick granular sub-base would enhance K from 0.054 to 0.073 which would reduce the stress to below the characteristic value.

4.8 Design method for point loads using Design Charts

The above example illustrates the complexity of undertaking calculations in the case of patch loads. The same method can be used for point loads (i.e. patch loads that approximate to a theoretical single point of application). In order to eliminate much of the effort, a series of design curves has been developed using Westergaard calculations. Although the Design Charts can be applied to patch loads, they are accurate only in the case of point loads such as the feet of storage systems. They can be used as described in the following.

(1) Assess the existing conditions: determine the actual point load (*APL*) and modulus of subgrade reaction (K) values from Section 4.4 to confirm the category of subgrade.

(2) Calculate the additional stress generated by point loads or wheels in close proximity.

- (i) If the distance between loads (S) is greater than 3 m, the APL can be used directly (depending on the radius of contact zone) to calculate the thickness of the slab using the relevant Design Chart. In this case, go directly to Stage 6.
- (ii) If the distance between loads is less than 3 m, the radius of relative stiffness (l) has to be determined. Table 4.3 shows values of radius of relative stiffness l for different K values and slab thicknesses.

Table 4.3. Radius of relative stiffness values (l) for different slab thicknesses and support conditions

Slab thickness: mm	Modulus of subgrade reaction K: N/mm^3			
	0.013	0.027	0.054	0.082
150	816	679	571	515
175	916	763	641	578
200	1012	843	709	639
225	1106	921	774	698
250	1196	997	838	755
275	1285	1071	900	811
300	1372	1143	961	865

From Figure 4.1, determine M_t/P, the ratio of the tangential moment to the greater point load by calculating S/l. Then use Equation (4.12) to determine the stress to add.

(3) From Table 4.1, select a proposed concrete mix and hence characteristic strength, σ_{char}.

(4) When two point loads are acting in close proximity (i.e. less than 3 m apart), the greater point load (P) produces a flexural strength σ_{max} directly beneath its point of application. The nearby smaller point load (P_2) produces additional stress σ_{add} beneath the larger load. Calculate σ_{max} from

$$\sigma_{max} = \sigma_{char} - \sigma_{add} \tag{4.13}$$

(5) Calculate the Single Point Load (SPL) which, acting alone would generate the same flexural stress as the actual loading configuration:

$$SPL = APL(\sigma_{char}/\sigma_{max}) \tag{4.14}$$

where APL is the actual point load.

(6) Prior to using the Design Charts, it is necessary to modify the SPL to account for contact area as well as wheel proximity to obtain the ESPL (Equivalent Single Point Load). Design Charts 1 to 10 apply directly when point loads have a radius of contact between 150 mm and 250 mm. Some racking systems and pallet transporters have a contact radius of less than 150 mm and some vehicles have a contact radius greater than 250 mm. In these cases multiply the point load by a factor in Table 4.4 prior to use in the Design Chart.

(7) Use the corresponding Design Chart for the mix selected in (3) to determine slab thickness and return to (3) if an alternative concrete mix is required.

Table 4.4. Point load multiplication factors for different values of K and for loads with a radius of contact outside the range 150 mm to 250 mm

Radius of contact: mm	Modulus of subgrade reaction K: N/mm^3			
	0.013	0.027	0.054	0.082
50	1.5	1.6	1.7	1.7
100	1.2	1.2	1.3	1.3
150	1.0	1.0	1.0	1.0
200	1.0	1.0	1.0	1.0
250	1.0	1.0	1.0	1.0
300	0.9	0.9	0.9	0.9

4.8.1. Design example for multiple point loading using Design Chart

A concrete floor is subjected to two point loads. A 60 kN point load is applied 1 m away from a 50 kN point load. The 60 kN point load has a contact zone radius of 100 mm and the 50 kN point load has a 300 mm radius. The existing ground conditions are poor (K = 0.027 N/mm^3). Assume the thickness of the slab to be 225 mm. The radius of relative stiffness l is given in Table 4.3 as l = 921 mm. The separation is 1 m (i.e. S = 1000 mm), so

$$\frac{S}{l} = \frac{1000}{921} = 1.086.$$

From Figure 4.1

$$M_t/P = 0.053.$$

Thus from Equation (4.12)

$$\sigma_{add} = 0.053\left(\frac{6}{200^2}\right)50\,000$$

$$\sigma_{add} = 0.4 \text{ N/mm}^2.$$

Try steel fibre reinforced C30 concrete with a characteristic strength of 2.2 N/mm^2 (30 kg/m^3 steel fibre dosage (see Table 4.1))

i.e.

$$\sigma_{char} = 2.2 \text{ N/mm}^2$$

$$\sigma_{\max} = \sigma_{\text{char}} - \sigma_{\text{add}}$$
$$= 2.2 - 0.4$$
$$= 1.8 \text{ N/mm}^2.$$

This is the maximum flexural stress which the 60 kN load can be allowed to develop. Calculate the *SPL* using Equation (4.14):

$$SPL = 60\left(\frac{2.2}{1.8}\right) = 7.3 \text{ kN}$$

From Table 4.4, the modified factor to be applied to the *SPL* to obtain the ESPL is 1.2:

$$73 \times 1.2 = 88 \text{ kN}$$

From Design Chart 4, the thickness of the slab = 225 mm.

Thus a 225 mm thick C30 concrete slab incorporating 30 kg/m^3 steel fibre is adequate for this design.

5. *Case studies*

5.1 Joint movement monitoring exercise

Because concrete is susceptible to the time-dependent behaviour of shrinkage, creep and thermal movement, joints are required in concrete ground floor slabs. These factors result in changes in internal stress if the slab is restrained from moving. The result, if left unchecked, would be shrinkage cracking and curling of the concrete slab. There are methods of predicting the shrinkage and creep of concrete[5.1] which are dependent on a number of variables.

5.1.1 Site investigation

A site investigation of joint performance was undertaken of two ground supported concrete industrial floors. Each 1000 m^2 floor was constructed using polypropylene fibre reinforced concrete by laser screeding. The two floors were Unit 114/14 and Unit 114/10 of the Boldon Business Park, Tyne & Wear, UK. Once the floor slabs had been concreted and the saw cut joints made, the arrangement of studs shown in Figs 5.1 and 5.2 was established on the concrete surface to permit joint movements to be monitored.

5.1.2 Site observations

Using an extensometer with an accuracy of one hundredth of a millimetre weekly measurements were taken over a period of nine months commencing November 1993.

5.1.3 Interpretation of results

Figures 5.3 to 5.12 show the movement of each of the measured joints from week to week. The larger movements at the beginning of the slabs' life represent joint cracking and the influence of cracked joints on neighbouring joints can be seen. When all the joints had cracked and were working, a more uniform movement became evident throughout the slab.

The data were used to calculate the cumulative movement of each joint with time in both floors. There are some sudden larger movements during the early stages of each slab's life which correspond to the active joint movement period. Once all the joints had cracked and were working, the movements became more uniform. The cumulative movements are of the same order, demonstrating uniform slab movement.

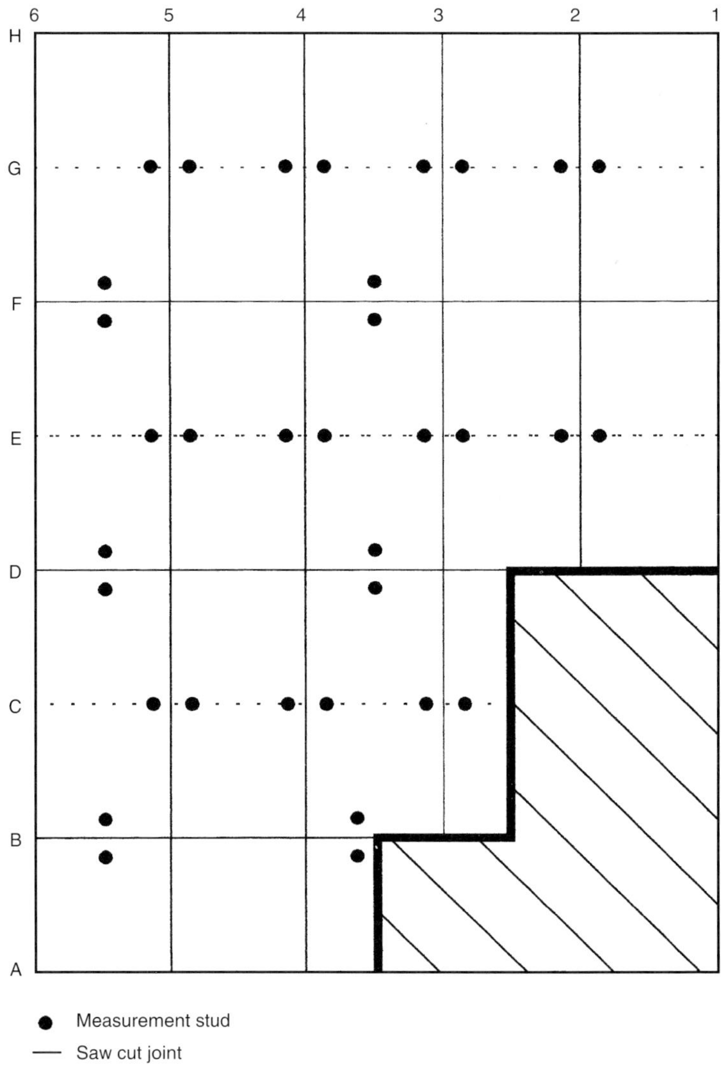

Fig. 5.1. Plan of joints in Unit 114/14

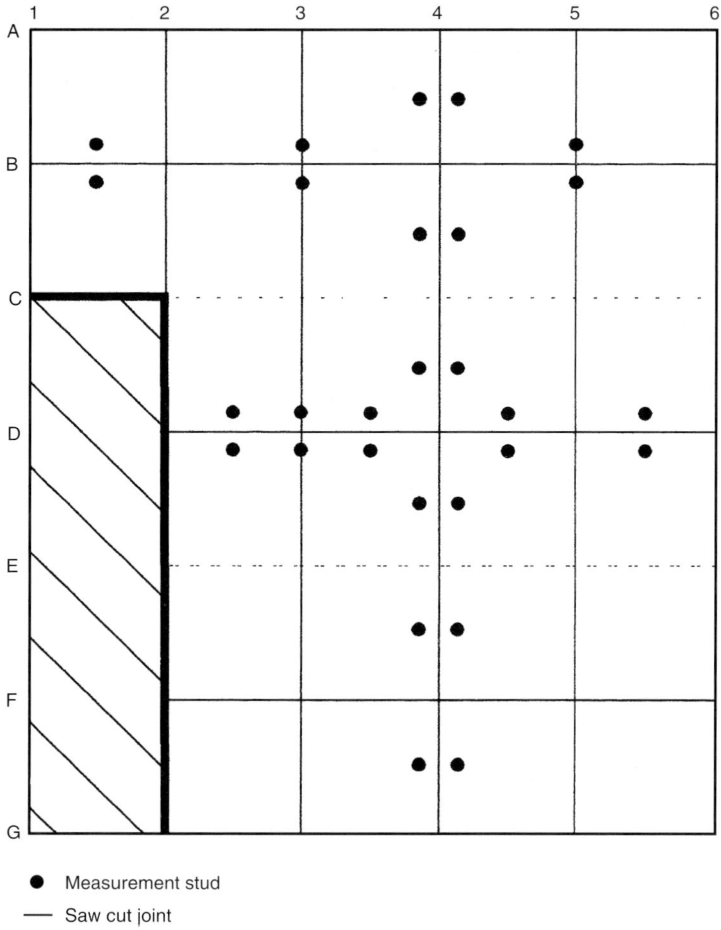

Fig. 5.2. Plan of joints in Unit 114/10

Figures 5.13 and 5.14 show the total movement observed at each location for both slabs. All of the joints opened gradually, and by the same degree. On average each joint opened by between 1 mm and 2 mm which is the optimum joint movement ensuring a working joint whilst at the same time maintaining aggregate interlock.

Monitoring of the two floors in this manner shows that the laser screeding process is capable of producing floors which behave in a way which should ensure successful performance in the long term. Monitoring of the two floors is continuing and ultimately a full life dataset will be available.

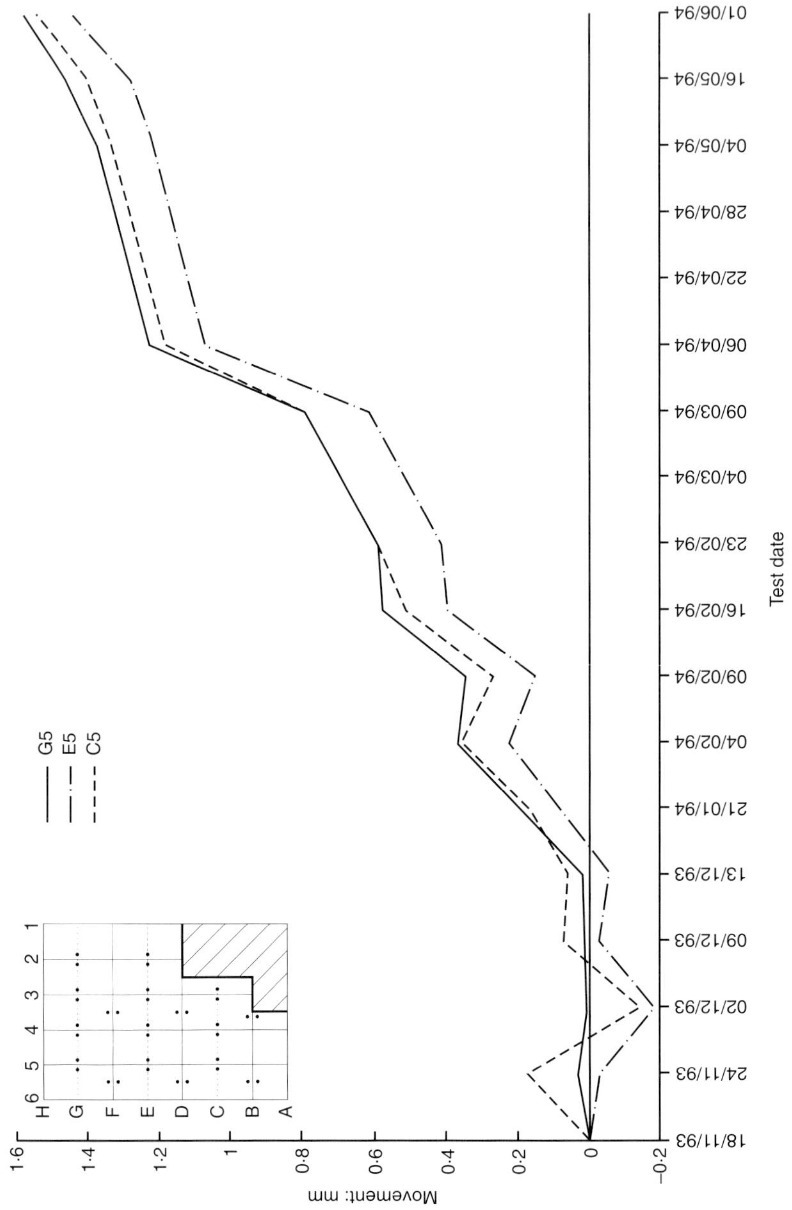

Fig. 5.3. Cumulative movements of joint 5 in Unit 114/14

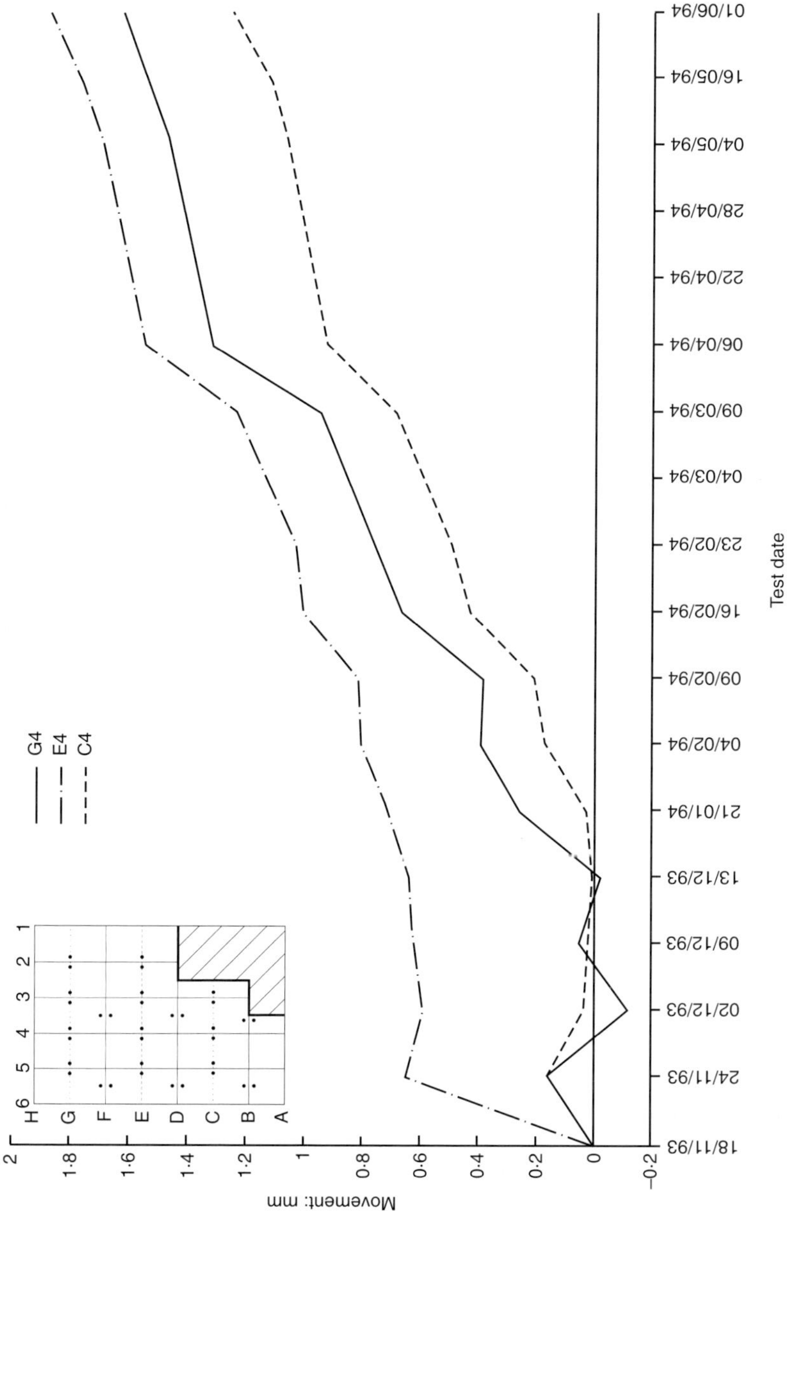

Fig. 5.4. Cumulative movements of joint 4 in Unit 114/14

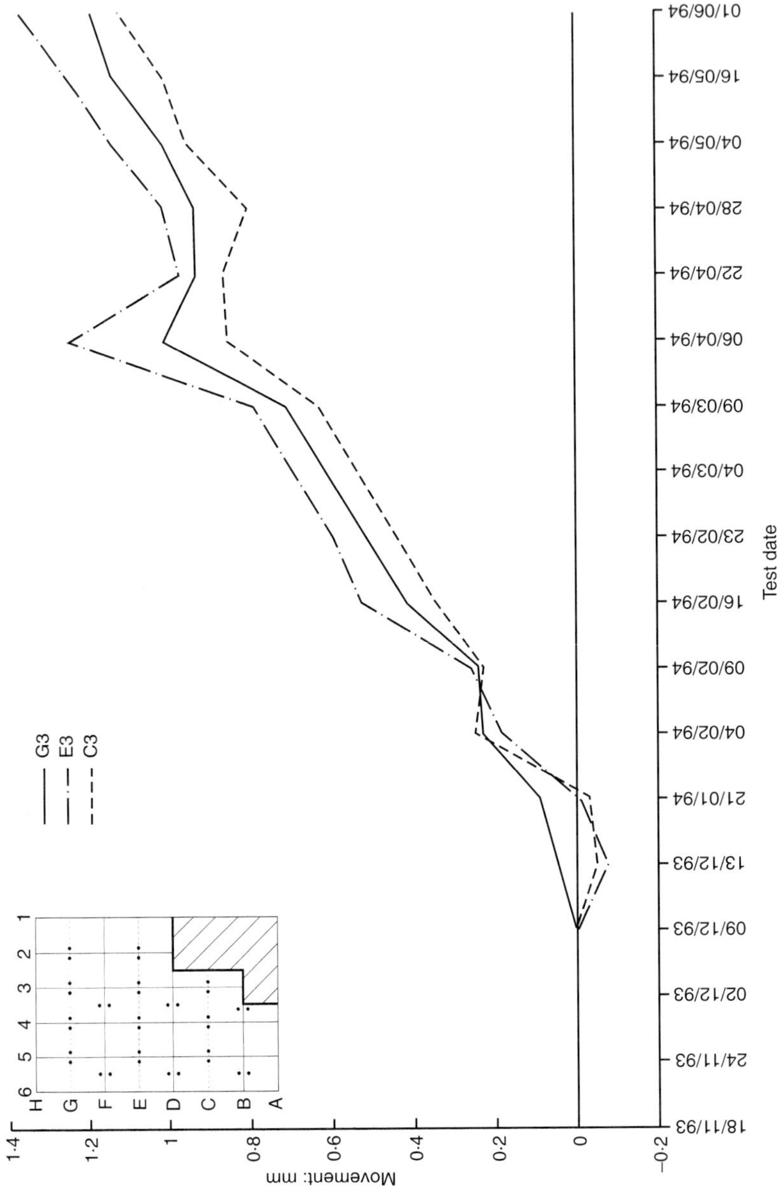

Fig. 5.5. Cumulative movements of joint 3 in Unit 114/14

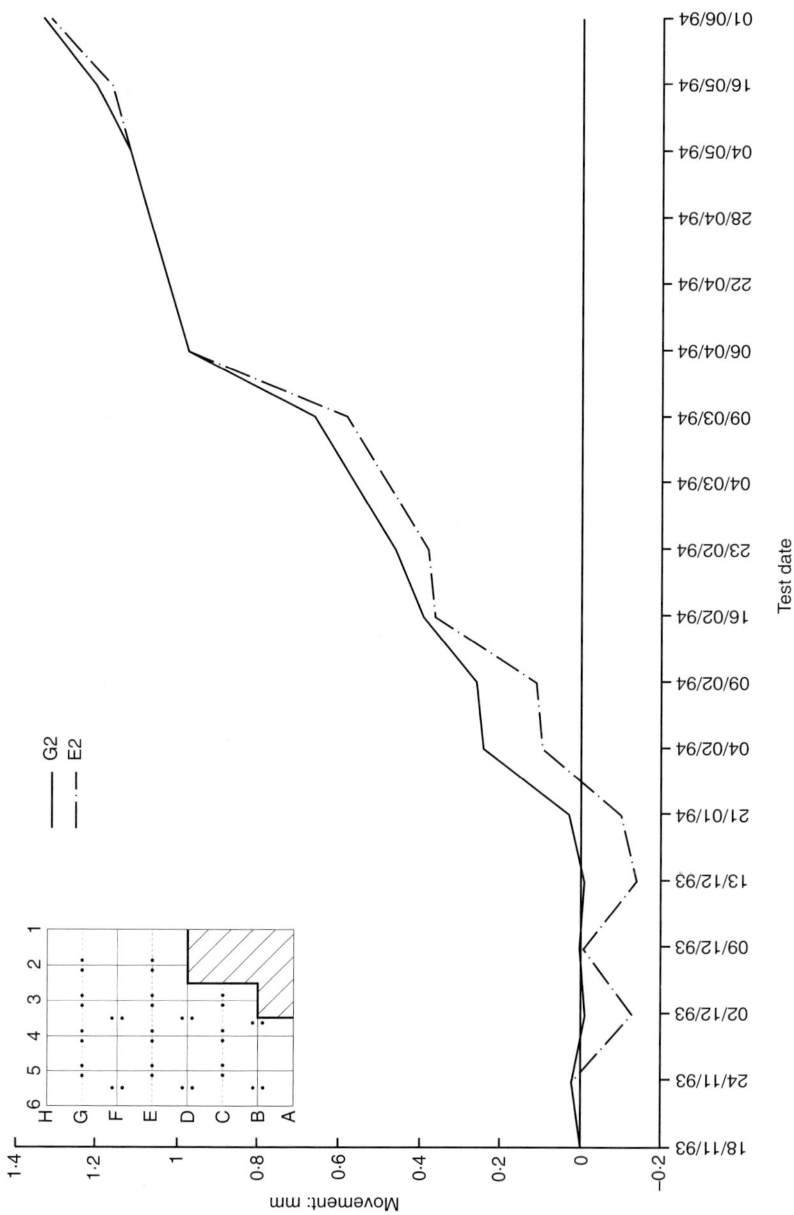

Fig. 5.6. Cumulative movements of joint 2 in Unit 114/14

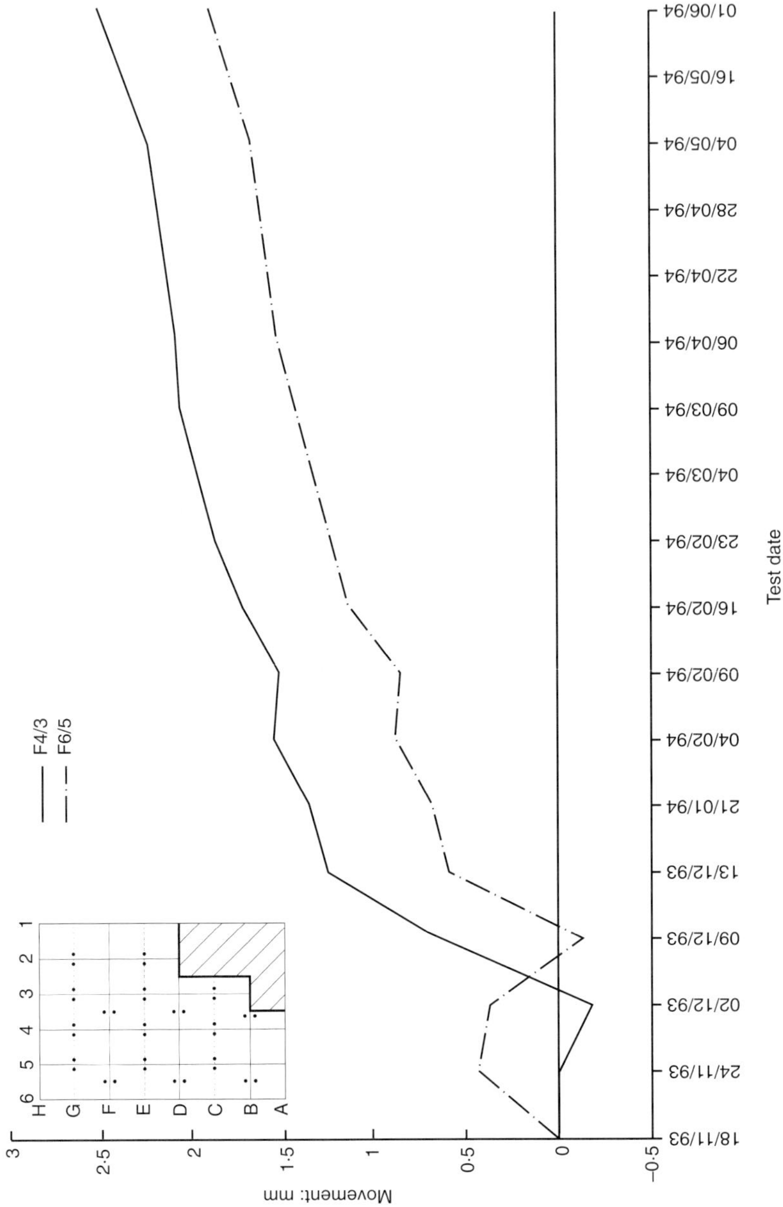

Fig. 5.7. Cumulative movements of joint F in Unit 114/14

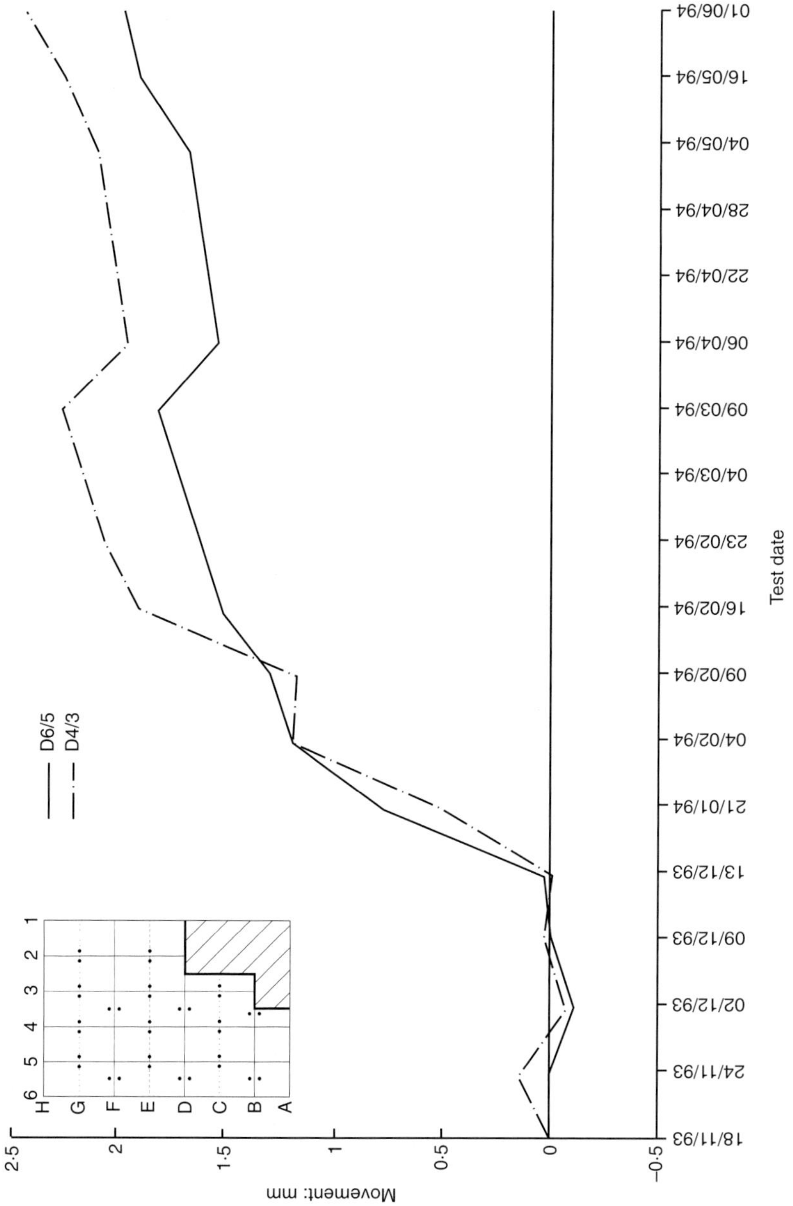

Fig. 5.8. Cumulative movements of joint D in Unit 114/14

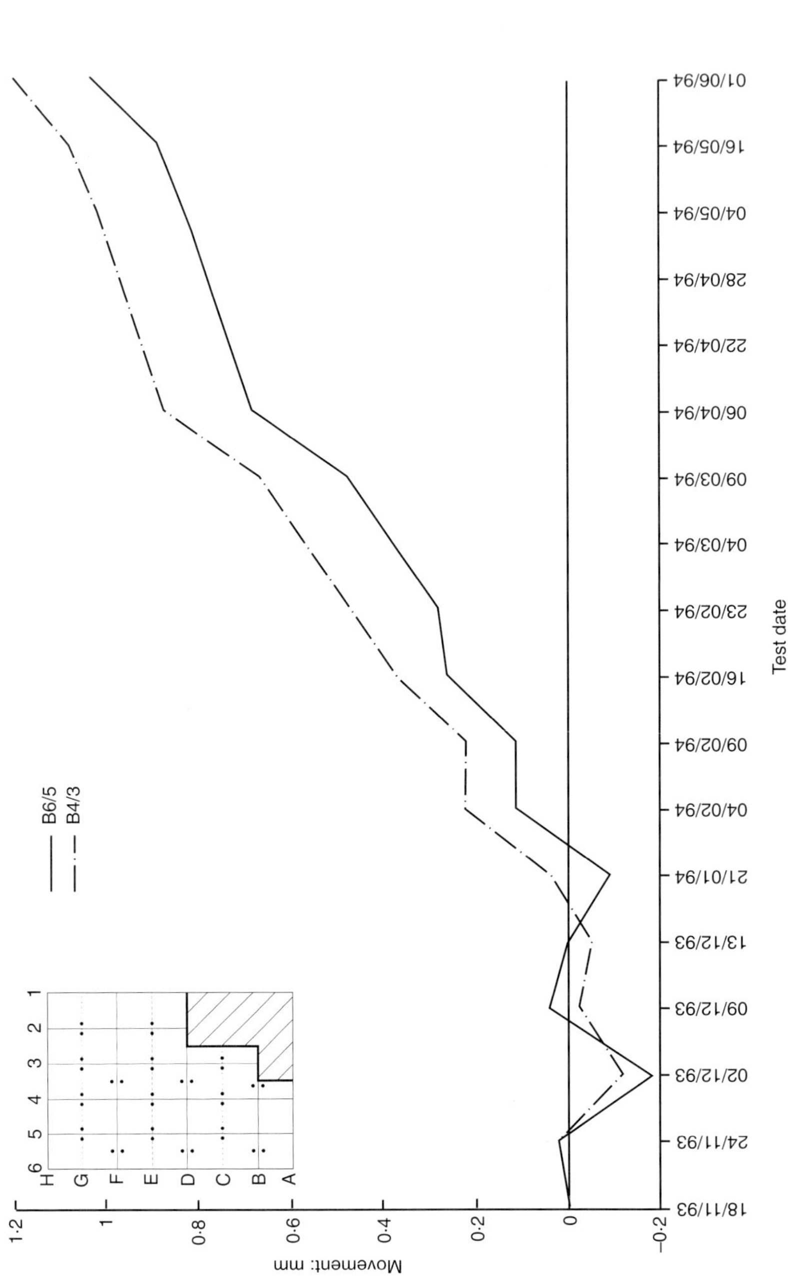

Fig. 5.9. Cumulative movements of joint B in Unit 114/14

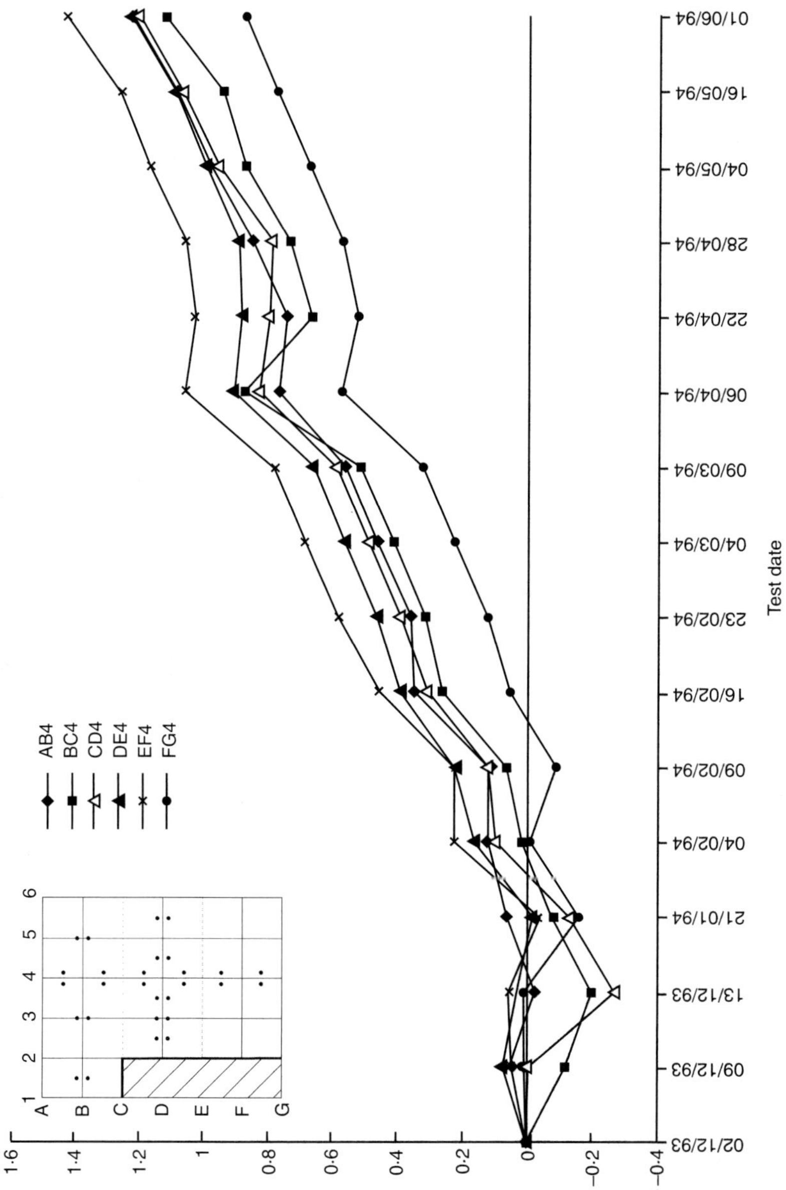

Fig. 5.10. Cumulative movements of joint 4 in Unit 114/10

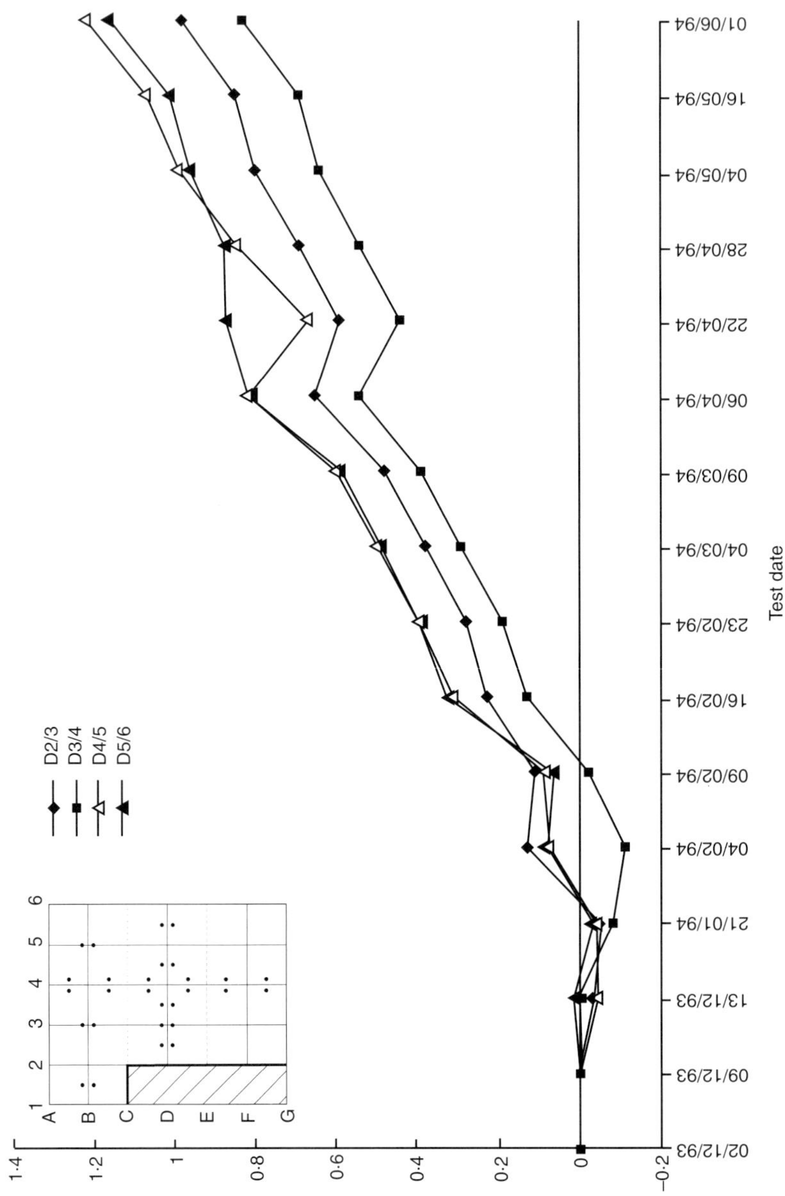

Fig. 5.11. Cumulative movements of joint D in Unit 114/10

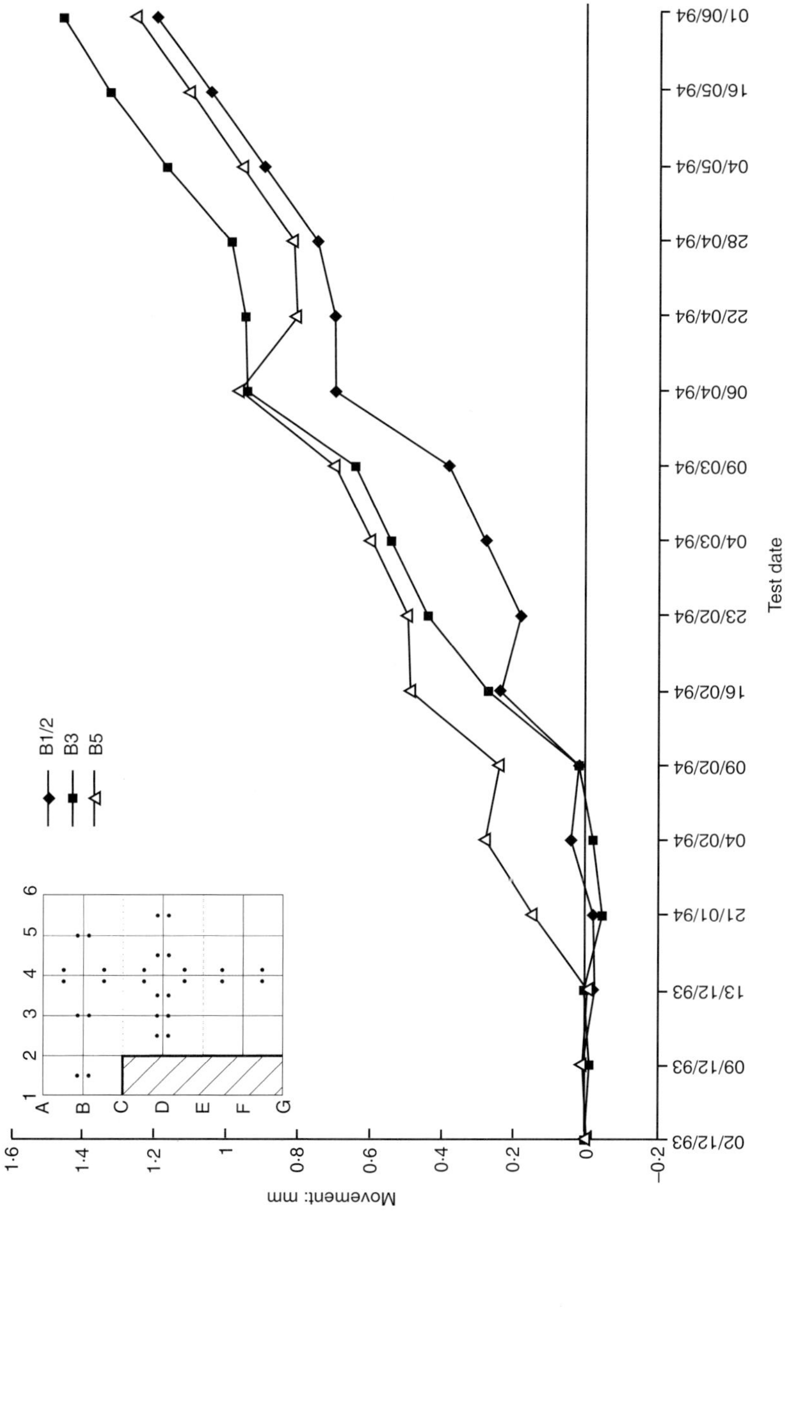

Fig. 5.12. Cumulative movements of joint B in Unit 114/10

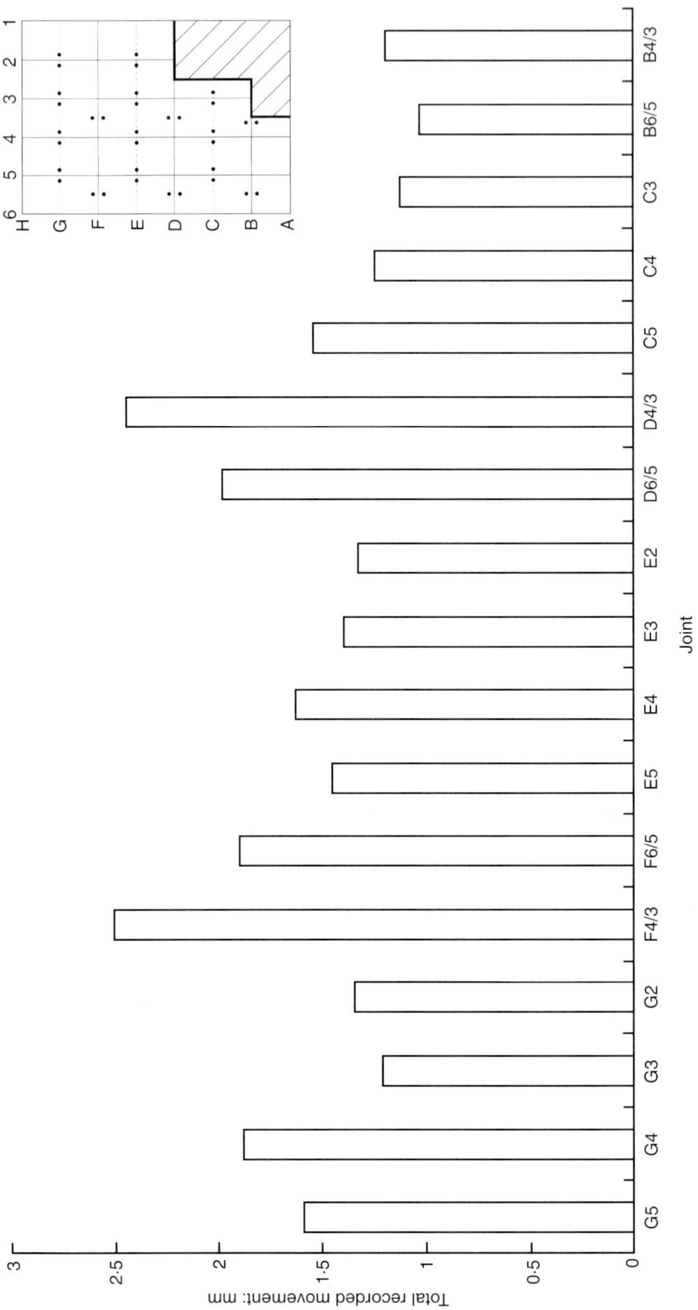

Fig. 5.13. Total recorded movements in Unit 114/14

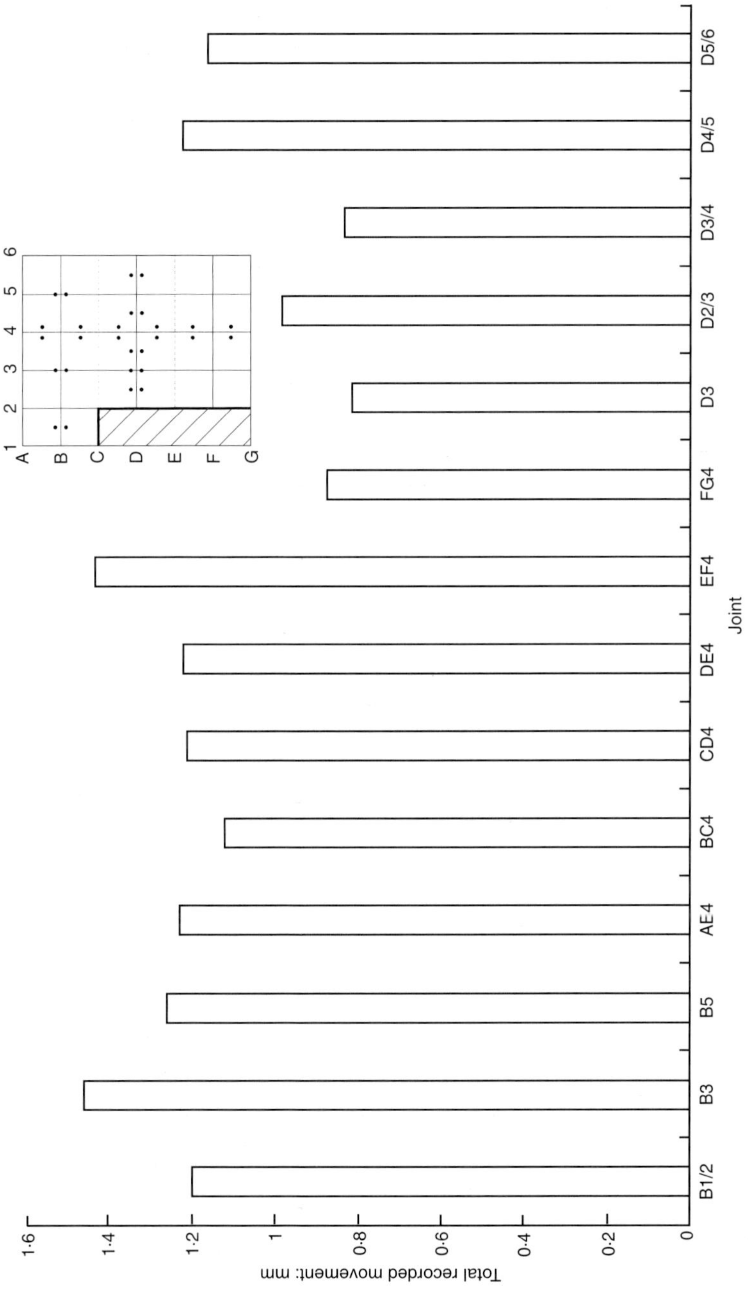

Fig. 5.14. Total recorded movements in Unit 114/10

The joint movements recorded have occurred as a result of drying shrinkage (temperature related shrinkage is usually of secondary importance in concrete floors). Three effects of drying shrinkage of concrete are friction, hogging and curling and these aspects are now considered in turn.

5.1.4 Friction

As a result of concrete shrinkage a concrete ground slab will shrink uniformly about its centre. Theoretically the centre will remain stationary. At a distance from the centre a horizontal displacement will occur and this displacement will increase uniformly towards the edge of the slab. The stress in the concrete resulting from frictional restraint to shrinkage can be calculated. The force required to overcome the friction force is given by the expression $F_f = w\mu$, where F_f is the friction force, w is the weight of the concrete (assumed to be 24 kN/m^3) and μ is the coefficient of friction. As the weight of concrete generating frictional restraint increases with distance from the slab edge the stress gradually increases to a maximum at the slab centre (there is zero stress at the edge of the slab or at the joints). Assuming the values for the coefficient of friction between concrete and sub-base and polythene are 0.65 and 0.15 respectively, the theoretical stresses which result from friction are shown in Figs 5.15 and 5.16.

5.1.5 Hogging

As a result of the underside of the concrete slab drying faster than the upper surface, non-uniform drying shrinkage develops throughout the slab with the lower zone of the concrete shrinking more than the upper. The result of this non-uniform shrinkage will be hogging of the slab. Assuming the hogging slab can be represented by a simply supported beam of length L with a uniformly distributed load equal to the concrete weight then the maximum moment, $M = wL^2/8$ where L is the length of the slab (the distance between joints) and w is the dead weight of the concrete assumed to be 24 kN/m^3. The stress is calculated from the equation $\sigma/y = M/I$ where σ is the stress, y is the depth to the neutral axis, M is the bending moment and I is the second moment of area. The relationship between slab length and hogging induced stress is shown in Fig. 5.17.

5.1.6 Curling

The result of the top of the slab drying faster than the bottom will cause the slab to attempt to curl upwards at its edges. This curling can be represented by a cantilever of length L, equal to the curled length.

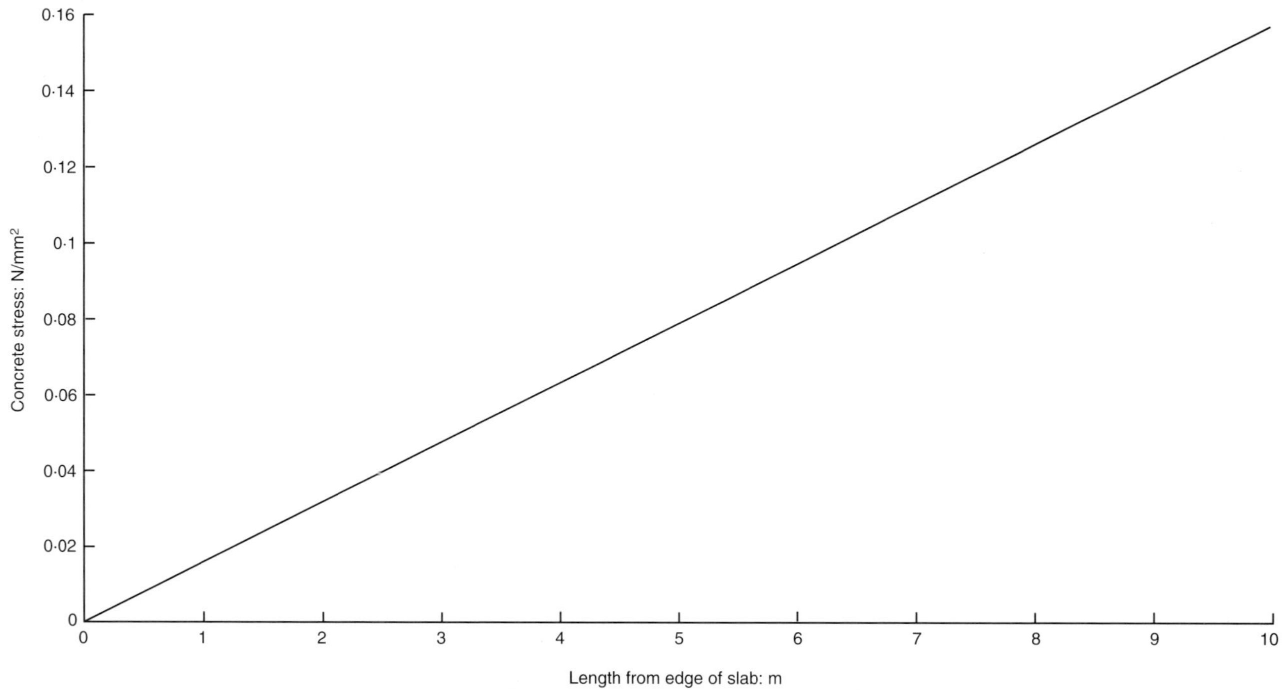

Fig. 5.15. Relationship between friction induced stress and distance from end of slab without slip membrane. Coefficient of friction $\mu = 0.65$ *(sub-base)*

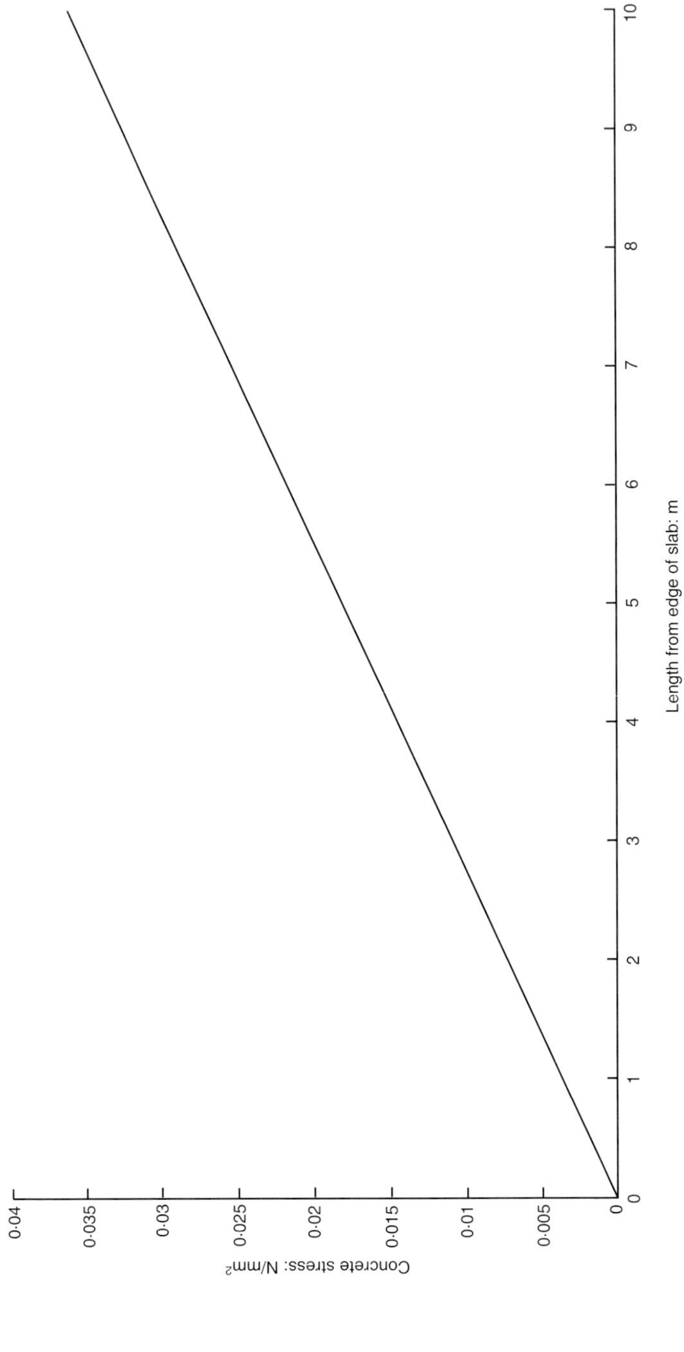

Fig. 5.16. Relationship between friction induced stress and distance from end of slab with slip membrane provided. Coefficient of friction $\mu = 0.15$ *(polythene)*

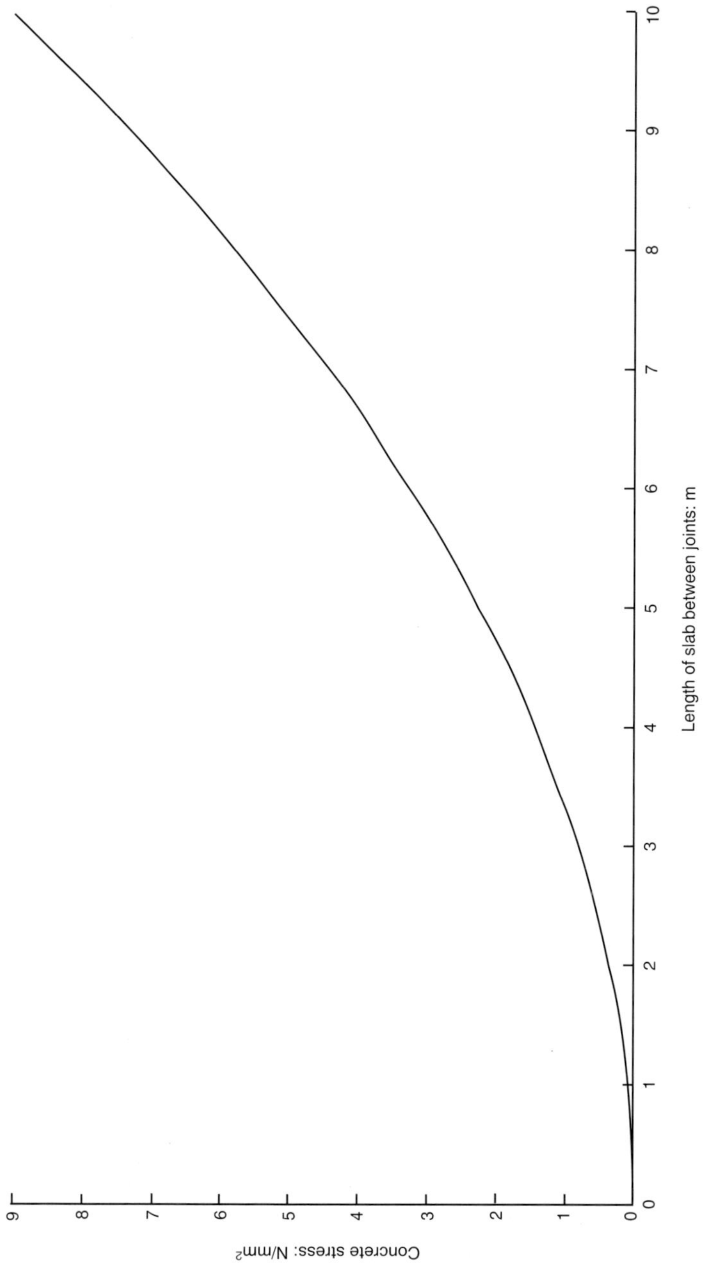

Fig. 5.17. Stresses in slab resulting from restraint to hogging

Assuming this cantilever carries a uniformly distributed load generated by the weight of the concrete (24 kN/m^3), the bending moment at the point of contact with the ground is given by the expression $M = wL^2/2$. As for hogging the stress can then be calculated from the expression $\sigma/y = M/I$; again the slab is assumed to have a breadth of 1 m and a depth of 200 mm. The relationship between slab length and curling induced stress is shown in Fig. 5.18.

5.1.7 Conclusion

The theoretical stresses induced by friction are negligible and the slip membrane is not required from the point of view of the stresses due to friction. However in the cases of hogging and curling the theoretical stresses can exceed the flexural strength of the concrete. Hogging is uncommon because the slip membrane reduces loss of moisture from the bottom of the slab so preventing it from curling. Good curing of the concrete will reduce drying shrinkage and provided the stresses develop slowly, concrete creep will reduce curling stresses to levels which require no additional reinforcement.

5.2 Design example

The following example illustrates the design process on a floor constructed using a laser guided screed. Figure 2.16 showed a plan of an industrial unit for which a ground bearing floor design was required. The floor is to be installed by a laser guided screeding machine as one pour. The internal walls are to constructed directly off the floor and the columns supporting the mezzanine floor are to be fixed onto the floor. The floor is to be constructed over poor ground with a modulus of subgrade reaction K of 0.027 N/mm^3. The floor area of 475 m^2 is well within the daily capacity of a laser screed, so it will be unneccesary to provide construction joints.

The floor is to comprise a steel fibre reinforced concrete slab constructed over a granular sub-base. The joint spacings, slab thickness, slab strength, fibre dosage and sub-base thickness are determined as follows.

The first consideration is joint spacing. A 600 mm perimeter beam will be cast around the perimeter of the building and a full movement joint will be constructed separating the floor slab from this beam. The only joints within the body of the floor slab will be induced joints formed by saw cutting 40 mm deep slots into the floor as soon as the concrete has gained sufficient strength to permit sawing without damaging the concrete. It is good practice for both aesthetic and structural reasons to align joints with column centres. In this case, joints could be provided on

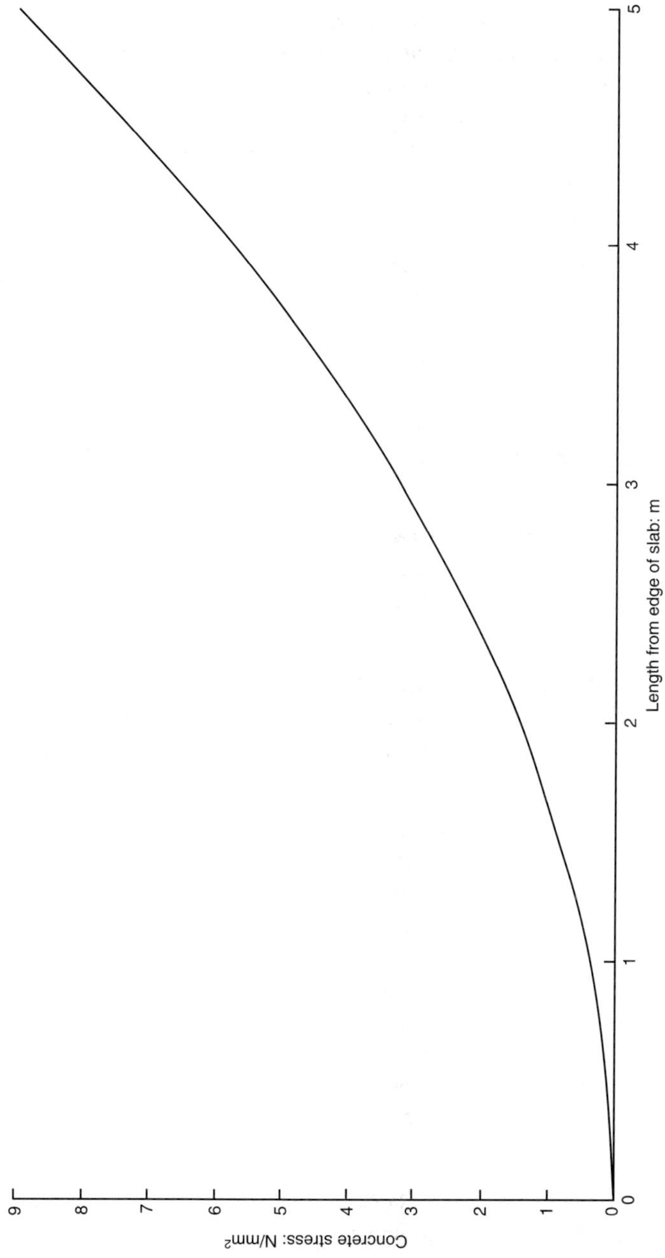

Fig. 5.18. Stresses in slab resulting from restraint to curling

grid lines 2, 3, 4 and B or alternatively on 3 and B. By providing the additional joints on grid lines 2 and 4, the maximum joint spacing is reduced from 12 m to 9 m. Reducing joint spacing allows a greater range of flooring materials to be considered. A third alternative, and the one which will be developed henceforth, is to provide joints on gridlines 2, 3, 4 and B and to provide additional joints half way between A–B and B–C. This reduces the maximum distance between joints to 6 m and has the benefit of providing bays which are nearly square, which is the most structurally efficient approach.

Table 2.1, reproduced here, shows that all of the commonly specified concretes are suitable.

A 20 kg/m^3 steel fibre C40 concrete will now be investigated. Table 4.1, reproduced here, shows that this fibre reinforced concrete has a characteristic flexural strength of 2.2 N/mm^2.

In order to account for fatigue, the three categories of loading are to be increased by the following factors:

mezzanine floor leg loads, $44 \times 1.5 = 66$ kN;
vehicle wheel load, $45 \times 2 = 90$ kN;
material storage load, $60 \times 1.25 = 75$ kN/m^2.

Table 2.1. Common concretes and suggested joint spacings.

Concrete type	Joint spacing: m
Plain C30 concrete	6
Micro-silica C30 concrete	6
20 kg/m^3 ZC 60/1.00 steel fibre reinforcement C30 concrete	6
30 kg/m^3 ZC 60/1.00 steel fibre reinforcement C30 concrete	8
40 kg/m^3 ZC 60/1.00 steel fibre reinforcement C30 concrete	10
Plain C40 concrete	6
Micro-silica C40 concrete	6
20 kg/m^3 ZC 60/1.00 steel fibre reinforcement C40 concrete	6
20 kg/m^3 ZC 60/1.00 steel fibre reinforcement C40 concrete	10
20 kg/m^3 ZC 60/1.00 steel fibre reinforcement C40 concrete	12

Table 4.1. Mean and characteristic 28 day flexural strength values for various concrete mixes

	Flexural strength: N/mm^2	
	Mean	Characteristic
Plain C30 concrete	2.0	1.4
Micro-silica C30 concrete	2.4	1.7
C30 concrete 20 kg/m^3 ZC 60/1.00 steel fibre[a]	2.8	2.0
C30 concrete 30 kg/m^3 ZC 60/1.00 steel fibre	3.2	2.2
C30 concrete 40 kg/m^3 ZC 60/1.00 steel fibre	3.8	2.7
Plain C40 Concrete	2.4	1.7
Micro-silica C40 Concrete	2.8	2.0
C40 concrete 20 kg/m^3 ZC 60/1.00 steel fibre	3.2	2.2
C40 concrete 30 kg/m^3 ZC 60/1.00 steel fibre	3.6	2.5
C40 concrete 40 kg/m^3 ZC 60/1.00 steel fibre	4.2	3.2

[a] ZC 60/1.00 refers to a commonly used anchored bright wire fibre of total length 60 mm and wire diameter 1.00 mm

The above three factors (1.5, 2 and 1.25) are based upon conventional fatigue relationships. Table 5.1 shows how increasing the load by various factors will extend the life of the floor. The system presented here allows different categories of loads to have different fatigue factors.

In the case of the mezzanine leg loads, the effect of load proximity needs to be taken into account (it is assumed here that the vehicle tyres are sufficiently well separated for proximity to be ignored – this is not always the case). Figure 4.1 is used to determine the ratio M_t/P. To do this, first, calculate the radius of relative stiffness l from:

Table 5.1. Relationship between number of load repetitions and fatigue factor. By multiplying the actual load by the fatigue factor, a higher load is produced which can be used to ensure a floor of the specified life

Number of load repetitions	Fatigue factor
1	1
50	1.25
5000	1.5
50 000	1.75
Infinity	2

$$l = \left(\frac{Eh^3}{12(1 - \nu^2)K}\right)^{0.25} \tag{5.1}$$

Usually, the elastic modulus E of fibre reinforced concrete is taken to be 20 000 N/mm^2 and Poisson's ratio, ν, is 0.15. At this stage, it is necessary to assume a slab thickness with no real guidance. As a trial, consider 225 mm. Equation (5.1) can now be evaluated:

$$l = [20\,000 \times 225^3/12(1 - 0.0225) \times 0.027]^{1/4}$$

$$l = 921 \text{ mm.}$$

The distance between the mezzanine leg loads is 1500 mm so the ratio S/l is 1500/921 = 1.63 where the terms have the meanings defined in Chapter 4. From Fig. 4.1, $M_t/P = 0.028$.

Equation (5.2) can now be used to determine the additional stress caused by one leg beneath its neighbour:

$$\sigma_{\text{add}} = \frac{M_t}{P}\frac{6}{h^2}P_2. \tag{5.2}$$

In this case, each leg load is 66 kN, so

$$\sigma_{\text{add}} = 0.028\frac{6}{225^2}66\,000 = 0.22 \text{ N/mm}^2$$

This stress is in fact added twice because the leg loads are in line which means one of the inner legs is subjected to the proximity effect twice.

Now use equation (5.3):

$$\sigma_{\max} = \sigma_{\text{flex}} - \sigma_{\text{add}} \tag{5.3}$$

to give

$$\sigma_{\max} = 2.2 - 2 \times 0.22 = 1.76 \text{ N/mm}^2$$

Now use equation (5.4) to determine the Single Point Load (*SPL*) which in this case accounts for the proximity of two neighbouring legs as well as the fatigue effect:

$$SPL = APL(\sigma_{\text{flex}}/\sigma_{\max}) \tag{5.4}$$

$$SPL = 66(2.20/1.76) = 82.5 \text{ kN} \tag{5.5}$$

If the effective radius of contact of the leg loads falls outside the range 150 mm to 250 mm, Table 4.4 is used to adjust the *SPL* to obtain the Equivalent Single Point Load (ESPL). In this case, it is assumed that the ESPL and the *SPL* are similar.

Having determined the ESPL, Design Chart 8 can now be used to determine the thickness of the floor. Before using the design chart, the effect of the granular sub-base in enhancing the modulus of subgrade reaction *K* must be determined. Table 1.5, reproduced here, is used to do this.

In this case, the underlined value (0.044) applies for a 250 mm thick sub-base and a subgrade *K* value of 0.027 N/mm^3. Using this value with the ESPL of 66 kN, Design Chart 8 confirms that the assumed slab thickness of 225 mm is satisfactory in the case of the mezzanine floor leg loads. Design Chart 8 is reproduced here with the 82.5 kN value brought across to the 0.044 *K* value curve (0.044 is interpolated between the 0.027 and 0.054 curves).

Now consider the highway vehicle and calculate the stresses in the floor when subjected to a patch load of 90 kN.

The three Westergaard equations introduced in Section 4.7 are reproduced here for three patch loads.

(a) Patch load within slab:

$$\sigma_{\max} = \frac{0.275(1+\nu)}{h^2} P \log\left(\frac{0.36Eh^3}{Kb^4}\right) \tag{5.6}$$

(b) Patch load at edge of slab:

$$\sigma_{\max} = 0.529(1+0.54\nu)\frac{P}{h^2} \log\left(\frac{0.20Eh^3}{Kb^4}\right) \tag{5.7}$$

Table 1.5. Enhanced value of K when a sub-base is used

K value of subgrade alone: N/mm^3	Enhanced value of *K* when used in conjunction with:							
	Granular sub-base of thickness (in mm)				Cement-bound sub-base of thickness (in mm)			
	150	200	250	300	100	150	200	250
0.014	0.018	0.022	0.027	0.033	0.045	0.063	0.081	0.106
0.027	0.034	0.038	<u>0.044</u>	0.051	0.075	0.104	0.137	—
0.054	0.059	0.065	0.072	0.081	0.125	0.175		
0.082	0.089	0.096	0.105	0.114				

Design Chart 8
20 kg/m^3 ZC 60/1·00 steel fibre
C40 concrete

Flexural strength = 3·2 N/mm^2

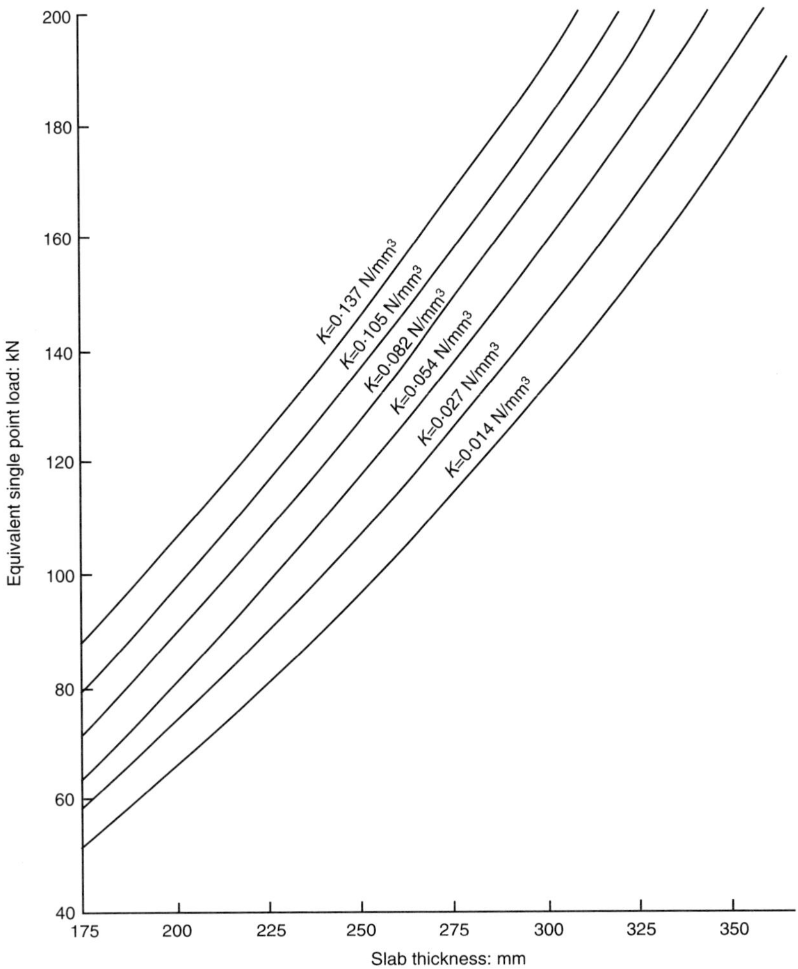

(c) Patch load at slab corner:

$$\sigma_{max} = \frac{3P}{h^2}\left[1 - \left(\frac{1.41b}{\{Eh^3/[12(1-\nu^2)K]\}^{0.25}}\right)^{0.6}\right] \tag{5.8}$$

where

σ_{max}	= flexural stress (in N/mm^2)
P	= patch load (in N) (i.e. characteristic wheel load × load factor, 90 000 N)
μ	= Poisson's ratio, 0.15
h	= slab thickness (in mm), 225 mm
E	= elastic modulus, 20 000 N/mm^2
K	= modulus of subgrade reaction, 0.044 N/mm^3
b	= radius of tyre contact zone (in mm) = $(P/\pi p)^{1/2}$
p	= contact stress between wheel and floor, 0.8 N/mm^2
b	= $(P/\pi p)^{1/2} - \sqrt{\frac{90\,000}{0.8\pi}} = 190$ mm.

Tables 5.2–5.7 can be used as an aid to evaluating the three Westergaard equations. They apply to the common μ and E values above. Westergaard stresses are obtained by multiplying together two values read from the appropriate tables.

5.2.1. Tables for patch load within slab

Table 5.2. Values for $[0.275(1+\nu)/h^2]P$. *The underlined figure relates to this example*

Load: N	Slab thickness: (in mm)								
	150	175	200	225	250	275	300	325	350
10 000	0.140	0.103	0.079	0.062	0.051	0.042	0.035	0.030	0.026
20 000	0.280	0.206	0.158	0.124	0.102	0.084	0.070	0.060	0.052
30 000	0.420	0.309	0.237	0.186	0.153	0.126	0.105	0.090	0.078
40 000	0.562	0.413	0.316	0.250	0.202	0.167	0.140	0.120	0.103
50 000	0.703	0.516	0.395	0.312	0.253	0.209	0.176	0.150	0.129
60 000	0.840	0.618	0.474	0.375	0.304	0.251	0.211	0.180	0.155
70 000	0.984	0.723	0.553	0.437	0.354	0.293	0.246	0.210	0.181
80 000	1.124	0.826	0.632	0.500	0.404	0.334	0.280	0.240	0.206
90 000	1.265	0.929	0.712	<u>0.562</u>	0.455	0.376	0.316	0.269	0.232
100 000	1.406	1.032	0.790	0.624	0.506	0.418	0.352	0.300	0.258
150 000	2.109	1.548	1.185	0.936	0.759	0.627	0.487	0.450	0.387
200 000	2.812	2.064	1.580	1.248	1.012	0.836	0.704	0.600	0.516

Table 5.3. Values for $\log(0.36Eh^3/Kb^4)$. *The underlined figure relates to this example*

Load: N	Slab thickness (in mm)								
	150	175	200	225	250	275	300	325	350
10 000	4.544	4.746	4.920	5.073	5.210	5.334	5.447	5.552	5.648
20 000	3.944	4.145	4.319	4.472	4.610	4.734	4.847	4.952	5.048
30 000	3.592	3.793	3.967	4.121	4.258	4.382	4.495	4.600	4.696
40 000	3.341	3.541	3.715	3.869	4.006	4.130	4.244	4.348	4.444
50 000	3.145	3.346	3.520	3.673	3.811	3.935	4.048	4.153	4.249
60 000	2.992	3.193	3.366	3.520	3.657	3.782	3.895	4.000	4.096
70 000	2.831	3.031	3.205	3.359	3.496	3.620	3.733	3.838	3.934
80 000	2.740	2.941	3.115	3.269	3.406	3.530	3.643	3.748	3.844
90 000	2.627	2.825	3.002	<u>3.155</u>	3.292	3.417	3.530	3.634	3.731
100 000	2.538	2.738	2.913	3.714	3.203	3.327	3.441	3.545	3.642
150 000	2.193	2.393	2.567	2.721	2.858	2.982	3.096	3.199	3.296
200 000	1.941	2.142	2.316	2.469	2.607	2.731	2.844	2.948	3.045

The figures in this table apply in the case of poor subgrade (K = 0.027 N/mm^3) with a 250 mm thick granular sub-base so the effective K value is 0.044 N/mm^3. In the case of very poor subgrade on 150 mm granular sub-base material, increase the values in the table by 12%. For floors constructed over good ground with 150 mm thickness of sub-base, decrease the values by 6%.

5.2.2. Tables for patch load at edge of slab

Table 5.4. Values for $[0.529(1 + 0.54\nu)/h^2]P$. *The underlined figure relates to this example*

Load: N	Slab thickness (in mm)								
	150	175	200	225	250	275	300	325	350
10 000	0.254	0.187	0.143	0.113	0.091	0.076	0.063	0.054	0.047
20 000	0.508	0.374	0.286	0.226	0.182	0.152	0.126	0.108	0.094
30 000	0.762	0.560	0.429	0.339	0.275	0.227	0.191	0.162	0.140
40 000	1.016	0.748	0.572	0.452	0.364	0.304	0.252	0.216	0.188
50,000	1.270	0.933	0.715	0.565	0.457	0.378	0.317	0.271	0.233
60 000	1.524	1.120	0.858	0.678	0.550	0.454	0.382	0.324	0.280
70 000	1.780	1.307	1.000	0.790	0.640	0.529	0.444	0.379	0.327
80 000	2.212	1.496	1.144	0.904	0.728	0.608	0.504	0.432	0.376
90 000	2.287	1.680	1.287	<u>1.017</u>	0.823	0.681	0.572	0.487	0.420
100 000	2.540	1.866	1.430	1.130	0.914	0.756	0.634	0.542	0.466
150 000	3.810	2.799	2.145	1.695	1.371	1.134	0.951	0.813	0.699
200 000	5.080	3.732	2.860	2.260	1.828	1.512	1.268	1.084	0.932

Table 5.5. Values for $\log(0.20Eh^3/Kb^4)$. *The underlined figure relates to this example*

Load N	Slab thickness (in mm)								
	150	175	200	225	250	275	300	325	350
10 000	4.290	4.490	4.664	4.818	4.955	5.079	5.193	5.297	5.393
20 000	3.689	3.890	4.064	4.217	4.355	4.479	4.592	4.697	4.793
30 000	3.337	3.538	3.712	3.865	4.003	4.127	4.240	4.344	4.441
40 000	3.085	3.286	3.460	3.613	3.751	3.875	3.988	4.093	4.189
50 000	2.890	3.091	3.265	3.418	3.555	3.679	3.793	3.897	3.994
60 000	2.737	2.938	3.111	3.265	3.402	3.526	3.640	3.744	3.840
70 000	2.575	2.776	2.950	3.104	3.241	3.365	3.478	4.071	3.679
80 000	2.250	2.686	2.860	3.013	3.151	3.275	3.388	3.492	3.589
90 000	2.372	2.572	2.747	<u>2.900</u>	3.037	3.162	3.275	3.379	3.475
100 000	2.282	2.483	2.658	2.811	2.948	3.072	3.186	3.290	3.387
150 000	1.937	2.138	2.312	2.466	2.603	2.727	2.840	2.944	3.041
200 000	1.686	1.886	2.061	2.214	2.351	2.476	2.589	2.693	2.790

The figures in this table apply in the case of poor subgrade ($K = 0.027\,N/mm^3$) with a 250 mm thick granular sub-base so the effective K value is $0.044\,N/mm^3$. In the case of very poor subgrade on 150 mm granular sub-base material, increase the values in the table by 12%. For floors constructed over good ground with 150 mm thickness of sub-base, decrease the values by 6%.

5.2.3. Tables for patch load at corner of slab

Table 5.6. Values for $1 - \{1.41b/[Eh^3/12(1-\nu^2)K]^{0.25}\}^{0.6}$. *The underlined figure relates to this example*

Load N	Slab thickness (in mm)								
	150	175	200	225	250	275	300	325	350
10 000	0.682	0.704	0.721	0.735	0.748	0.758	0.768	0.776	0.783
20 000	0.609	0.636	0.657	0.674	0.689	0.703	0.714	0.724	0.733
30 000	0.559	0.588	0.612	0.632	0.649	0.664	0.677	0.688	0.699
40 000	0.519	0.551	0.577	0.599	0.618	0.634	0.648	0.660	0.671
50 000	0.485	0.520	0.548	0.571	0.591	0.608	0.624	0.636	0.648
60 000	0.457	0.494	0.523	0.547	0.568	0.586	0.603	0.617	0.629
70 000	0.426	0.464	0.496	0.522	0.544	0.563	0.580	0.594	0.608
80 000	0.407	0.448	0.480	0.507	0.529	0.549	0.560	0.582	0.596
90 000	0.384	0.426	0.459	<u>0.487</u>	0.511	0.531	0.550	0.565	0.579
100 000	0.365	0.407	0.442	0.471	0.495	0.516	0.536	0.551	0.566
150 000	0.284	0.333	0.371	0.404	0.432	0.455	0.477	0.495	0.511
200 000	0.219	0.272	0.314	0.350	0.380	0.406	0.429	0.449	0.467

Table 5.7. Values for $3P/h^2$. The underlined figure relates to this example

Load N	Slab thickness (in mm)								
	150	175	200	225	250	275	300	325	350
10 000	1.333	0.980	0.750	0.593	0.480	0.397	0.333	0.284	0.245
20 000	2.667	1.959	1.500	1.185	0.960	0.793	0.667	0.568	0.490
30 000	4.000	2.938	2.250	1.778	1.440	1.190	1.000	0.850	0.735
40 000	5.333	3.918	3.000	2.370	1.920	1.587	1.333	1.136	0.980
50 000	6.667	4.898	3.750	2.963	2.400	1.983	1.667	1.420	1.224
60 000	8.000	5.877	4.500	3.555	2.880	2.380	2.000	1.704	1.469
70 000	9.333	6.857	5.250	4.148	3.360	2.777	2.333	1.988	1.714
80 000	10.667	7.836	6.000	4.740	3.840	3.174	2.667	2.272	1.959
90 000	12.000	8.816	6.750	5.333	4.320	3.570	3.000	2.556	2.204
100 000	13.333	9.796	7.500	5.926	4.800	3.967	3.333	2.840	2.449
150 000	20.000	14.694	11.250	8.889	7.200	5.950	5.000	4.260	3.673
200 000	26.667	19.592	15.000	11.852	9.600	7.934	6.667	5.680	4.898

5.2.4. Solution

The stresses developed in the slab by a patch load of 90 kN can now be calculated using the values underlined in Tables 5.2–5.7 as follows.

(a) Maximum stress within the slab: $0.562 \times 3.155 = 1.77\ \text{N/mm}^2$.
(b) Maximum stress at slab edge: $1.017 \times 2.900 = 2.95\ \text{N/mm}^2$.
(c) Maximum stress at corner of slab: $0.487 \times 5.333 = 2.60\ \text{N/mm}^2$.

In the case of edge and corner stresses, aggregate interlock and fibre load transfer may allow the sharing of patch loads by neighbouring slabs. This transfer reduces edge and corner stresses by an amount which depends upon the amount by which the joints open. The true joint opening depends upon many factors, some of which will be unknown at the design stage. Experience indicates that significant load transfer occurs when joints open by up to 1 mm. When joint openings exceed 2 mm, there is little or no load transfer. In the absence of other data, Table 5.8 may be used to assess load transfer. The table has been developed from the Author's experience in the design and maintenance of floors.

In this example, joints are at 6 m spacings in both directions, so the maximum 30% load transfer can be assumed. This applies to both edge and corner loading. Therefore, the three critical stresses are as follows.

(a) Maximum stress within the slab $= 1.77\ \text{N/mm}^2$.
(b) Maximum stress at slab edge $= 2.07\ \text{N/mm}^2$.
(c) Maximum stress at corner of slab $= 1.82\ \text{N/mm}^2$.

The characteristic strength of the concrete is $2.2\ \text{N/mm}^2$ so the proposed slab is satisfactory in the case of patch loading.

Now consider a materials storage load of $75\ \text{kN/m}^2$ or $0.075\ \text{N/mm}^2$. The stress produced by the most onerous combination of storage space and aisles is given by:

$$\sigma_{\max} = 1008q/(\lambda^2 h^2) \text{ (in N/mm}^2\text{)}. \quad (5.9)$$

Table 5.8. Dowel bar load transfer efficiency values

Joint spacing: m	Load transfer: %
6 or less	30
8	20
10	10
12	0

Table 4.2. Values of $\lambda^2 h^2$ for combinations of slab thickness and modulus of subgrade reaction

Modulus of subgrade reaction K: N/mm^3	$\lambda^2 h^2$ for slab thickness of (in mm):				
	150	175	200	225	250
0.082	0.061	0.066	0.070	0.074	0.078
0.054	0.049	0.053	0.057	0.060	0.063
0.044	0.043	0.047	0.050	0.053	0.056
0.027	0.035	0.038	0.040	0.043	0.045
0.013	0.024	0.026	0.028	0.029	0.031

Table 4.2, which shows values of $\lambda^2 h^2$ for common combinations of modulus of subgrade reaction (K) and slab thickness, is reproduced above with an additional line showing values for $K = 0.044\,\text{N/mm}^3$ which applies to this example. The appropriate value of $\lambda^2 h^2$ is shown underlined.

Substituting known values into the stress equation (Equation (5.9)) yields

$$\sigma_{\text{max}} = 1008 \times 0.075/0.053 = 1.426\ \text{N/mm}^2.$$

This is less than the characteristic strength of 2.2 N/mm^2 so the floor is satisfactory for the distributed load.

Figure 2.17 illustrates the solution.

5.3. Flatness

5.3.1. The need for a flat floor

Flatness is an essential requirement of many ground floor slabs especially in the aisles of very narrow aisle (VNA) high density warehouses where free path fork lift trucks operate. Variations in level across the aisle between the wheel tracks of the truck are magnified in proportion to the ratio of the height of the racking to the trackwidth of the truck. It is not uncommon for such vehicles to collide with high racking if one wheel is at a different level from the equivalent wheel on the other side. Warehouses are designed to achieve a throughput of a given number of pallets per hour and a poor floor may reduce the number of pallets handled. Furthermore, storage systems which have non-adjustable feet require particularly flat floors. Other situations where floor flatness is an important property include those trafficked by automatically controlled transportation equipment and the movement of heavy plant within a factory by hover transportation.

Floors are divided into those where movement is defined and those allowing free movement. Defined movement areas include the aisles in VNA warehouses whereas free movement areas include warehouses with wide aisles or free running areas. Defined movement areas need to have enhanced flatness characteristics as compared with free movement areas. The criteria defined in Concrete Society publication TR34[1.2] are shown in Tables 5.9 and 5.10. The floor specifier should define the category of flatness required.

5.3.2. Flatness of laser screeded floors

A floor with a very high degree of flatness and levelness can be produced using a laser screed. In most cases, a good Category 2 tolerance will be achieved. With careful supervision an experienced laser screed contractor will be able to achieve a Category 1 floor but in some cases it may be necessary to undertake some remedial grinding in the wheelpaths of lift trucks within racking aisles.

5.4 Engineering data for fibre reinforced concrete

The following test data have been obtained as the use of steel and polypropylene fibres has increased.

5.4.1. Data relating to steel fibre reinforced concrete

Several research programmes have been undertaken on slabs incorporating steel fibre reinforced concrete. Two are summarized below.

(a) Imperial College, London (Table 5.11)
- 12 slabs tested (1 m × 1 m × 50 mm)
- Simply supported on four edges and loaded in the centre
- Same basic concrete mixture
- Slabs contain varying quantity of hooked 60 mm long, 1.0 mm diameter fibres

(b) Thames Polytechnic, London (Table 5.12)
- 9 slabs tested (3 m × 3 m × 150 mm)
- $K = 0.35$ N/mm[3.]
- Loaded until failure with point load in centre of slab
- Load at first visible crack also recorded

5.4.2. Data relating to polypropylene fibre reinforced concrete

The data that follow were obtained at Signet Laboratory and at San Jose State University using plain concrete and concrete with Fibermesh polypropylene fibres added. The concrete had a cement content of

Table 5.9. Flatness requirements for defined movement areas

Category	Location	Allowable limits: mm							
		Property I		Property II		Property III			
		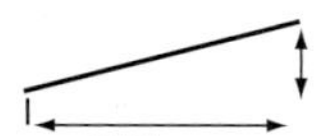 I 300 mm		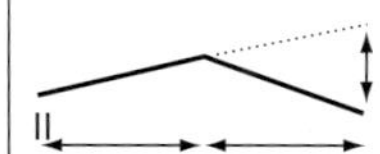 II 300 mm 300 mm		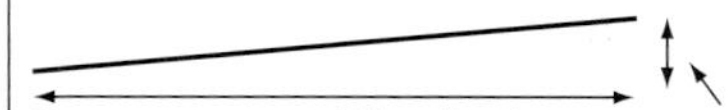 III Wheeltrack			
						Wheeltrack up to 1.5 m		Wheeltrack over 1.5 m	
		A	B	A	B	A	B	A	B
Superflat (SF)	VNA warehouses with minimum clearance between fixed and moving pallets. Maximum throughputs, truck speed and permitted rack height.	0.75	1.00	1.00	1.50	1.50	2.50	2.00	3.00
Category 1	VNA warehouses with racking height between 8 and 13 m. Top guided trucks between 13 and 20 m.	1.50	2.50	2.50	3.50	2.50	3.50	3.00	4.50
Category 2	VNA warehouses with racking height less than 8 m.	2.50	4.00	3.25	5.00	3.50	5.00	4.00	6.00

Notes:

Tolerance of level to datum plane (in mm): Category SF ±10
Category 1 ±10
Category 2 ±15

The floor can be considered satisfactory when:

(a) not more than 5% of the total number of measurements exceed the particular property limit in column A and
(b) none of the measurements exceed the particular property limit in column B.

Table 5.10. Flatness requirements for free movement areas

Floor classification	Location	Maximum permissible limits (mm)		
		Property II (as for Table 5.9)	Property IV: difference in elevation (in mm) of points on a 3 m grid	
			A	B
FM1	Hover transport and areas of special consideration	2.5	3.0	4.5
FM 2	Wide aisle warehouses, automatically guided transfer vehicles, transfer aisles	3.5	5.5	8.0
FM3	Wide aisle warehousing using counter balanced trucks. Manufacturing facility, general warehousing for block stacking.	5.0	5.5	8.0

Notes:
The floor can be considered satisfactory when:

(a) not more than 5% of the total number of measurements exceed the particular property limit in column A and
(b) none of the measurements exceed the particular property limit in column B or the property II column.

Table 5.11. Summary of results of work undertaken at Imperial College, London

Fibre dosage: kg/m^3	0	20	25	30
First-peak load: kip	2.09	2.14	2.47	2.47
First-peak stress: psi	638	652	754	754
Maximum load: kip	2.11	2.79	2.97	3.10
Plateau load: kip[a]	0	1.96	2.36	2.47

[a] plateau load corresponds to a state of increasing deflection under constant load and represents the collapse load of the slab. 1 kip = 1000 lbf.

Table 5.12. Summary of results of work undertaken at Thames Polytechnic, London

	Steel fibre type				
	None	Length 60 mm, diam. 1 mm	Length 60 mm diam. 1 mm	Length 60 mm diam. 0.8 mm	Length 60 mm diam. 0.8 mm
Quantity: kg/m^3	0	20	30	20	30
Load at first visible crack: kip	40.5	47.2	53.9	58.4	65.2
Maximum load: kip	45.0	73.1	76.4	87.7	77.6[a]

[a] In the first test the capacity of the testing apparatus was not sufficient; for subsequent tests the apparatus was modified.

240 kg/m^3, water/cement ratio of 0.64 and maximum aggregate size of 20 mm. The following results (Tables 5.13–5.15) were obtained.

5.4.2.1 Rate of gain of compressive strength. Table 5.13 shows the rate of gain of compressive strength for both plain and polypropylene reinforced concrete. Fibre was added at the rate of 890 g/m^3.

5.4.2.2 Effect of fibre dosage on compressive strength. Table 5.14 shows how three different levels of fibre dosage influence the compressive strength of concrete at age 7 and 28 days.

Table 5.13.

Age of concrete	Compressive strength: N/mm^2	
	Plain concrete	Polypropylene fibre reinforced concrete
18 h	2.5	2.6
24 h	3.3	3.4
7 days	9.8	10.4
28 days	16.0	17.4

5.4.2.3 Effect of fibre dosage on flexural strength and tensile strength. Table 5.15 shows flexural strength values determined by bending beams of dimensions 150 mm × 150 mm × 450 mm and tensile strength values obtained by undertaking cylinder splitting tests to ASTM C-490.

5.5. Long strip floor construction case study

This example of a floor constructed conventionally by the long strip method is presented in order to allow a comparison with the single pour method described in Chapter 2. The floor was constructed in 1998 at the Silverlink Industrial Estate in North Tyneside. A conventionally reinforced 150 mm thick concrete slab was constructed over a dolomitic limestone 150 mm thick sub-base. The floor was constructed over a period of two weeks. One or two strips running from one side of the building to the other were constructed each working day. The reinforcement was placed near the upper surface of the slab and three types of joint were provided. The joints between neighbouring strips comprised conventional ties connecting the strips. The transverse joints comprised dowelled joints and tied joints. In the case of the dowelled joints, sleeves were placed over the bars at one side of the joint to allow for contraction. The ties comprised a short length of mesh reinforcement placed at the underside of the slab.

Table 5.14.

Polypropylene fibre dosage: g/m^3	Compressive strength: N/mm^2	
	7 day strength	28 day strength
0	14.2	27.3
600	15.9	29.7
1200	16.2	30.4

Table 5.15.

Polypropylene fibre dosage: g/m^3	Flexural strength: N/mm^2		Tensile strength: N/mm^2	
	7 days	28 days	7 days	28 days
0	1.8	2.5	1.2	1.9
890	2.0	2.7	1.4	2.0

This case study is similar in some respects to the one described in Section 2.9, except no perimeter beam is included and the surrounds to the columns are square. Details of the project are included in the captions to Figs. 5.19 to 5.37.

Fig. 5.19. The propped portal frame has been completed and the sub-base has been laid. Floor construction cannot start until the roof sheets have been fixed otherwise the sun could lead to rapid moisture loss from the concrete. This in turn could cause plastic cracking in the floor

Fig. 5.20. The prop columns along the centre of the building have been bolted to their concrete foundations which are below the level of the underside of the floor slab. The floor slab will surround these columns

Fig. 5.21. Lightweight steel mesh reinforcement has been fixed around the prop columns through the depth of the floor slab. A baseplate has been welded to the end of the column. The baseplate has been bolted to the foundation. Wind may cause the column to develop tension so it has to be bolted positively to its foundation

Fig. 5.22. Concrete has been cast around the column through the depth of the floor slab. A strip of compressible material will be fixed around the perimeter of this concrete to the full depth of the floor to ensure that the floor and the structural frame can move independently of each other

Fig. 5.23. A slip membrane has been laid and formwork has been fixed prior to concreting. The concrete truck reverses along the strip to the discharge position. Care has to be taken to avoid rutting the sub-base and disturbing the slip membrane. The mesh reinforcement cannot be installed prior to concreting as it would interfere with access

Fig. 5.24. The concrete is discharged between the formwork and, in this case, a previously constructed strip. A piece of mesh reinforcement has been installed. The concrete is discharged as close as possible to its required position and is moved by hand as required

Fig. 5.25. A standard size (4.8 m × 2.4 m) piece of mesh is ready to be placed in the fresh concrete. The mesh is orientated so the main reinforcement is in the longitudinal direction. The mesh is positioned vertically, then allowed to rotate onto the surface of the concrete

Fig. 5.26. The mesh is pressed into the concrete by members of the concrete gang walking over it. The mesh can be located with a surprising degree of accuracy by a skilled team. Care is taken to ensure sufficient overlap with neighbouring pieces of mesh. A cropping tool is on hand to deal with obstructions

Fig. 5.27. A double vibrating beam is pulled along the formwork to compact the concrete. A member of the team ensures that there is sufficient concrete along the length of the beam. Tie bars have been inserted in holes in the formwork

Fig. 5.28. A short length of mesh is placed near the underside of the concrete to act as the tie at a transverse tied joint. The main mesh reinforcement near the upper surface of the concrete is stopped short of the tied joint. Note that the main reinforcement runs normal to the longitudinal long strip direction. A transverse groove has already been cut in the neighbouring bay to complete the tied joint

Fig. 5.29. The edges of the slab are trowelled to improve the quality of the longitudinal joint. This trowelling takes place immediately after the concrete has been compacted. Additional concrete may be placed to make up levels

Figure 5.30. The concrete is left to set. Note the excess slip membrane material and the tie bars. Because the roof sheets are fixed and the weather is cool, there is no need to take special measures to reduce moisture loss from the upper surface of the concrete

Fig. 5.31. The floor slab has been cast around an internal column. Compressible material has been fixed around the square concrete surrounding the column. The upper part of the compressible material will be replaced by a protective strip

Fig. 5.32. These powertrowels are used to improve the quality of the surface of the concrete. They are used when the concrete has gained sufficient strength to support their weight and the weight of the operator, but before the concrete is fully hard. In this case, concrete mixed at 9a.m. was powertrowelled at 1p.m.

Fig. 5.33. The effect of powertrowelling is to create a smooth and durable surface. There must be just the correct amount of surface laitance. Too little and the surface will remain open; too much and a thin layer may eventually spall at the surface

Fig. 5.34. The finished slab with the formwork removed showing the tie bars ready for the neighbouring long strip. This concrete can be used as formwork for the next long strip. Sometimes the alternate strip method is used whereby the initial long strips require formwork at each side and the intermediate strips use the previously cast concrete. In this case, one central long strip was cast using formwork at each side. All of the subsequent strips were cast using one existing concrete long strip and one side of formwork

Fig. 5.35. An isolation joint is formed at the building perimeter using the compressible material shown. In this case, the floor slab is constructed right up to the internal wall without a perimeter strip. The slip membrane is brought to the slab surface to contain the moisture in the concrete during curing

Fig. 5.36. At the external columns, the concrete block internal leaf is constructed round each column and a compressible strip is fixed to the concrete blocks prior to concreting. This isolates the floor from the internal leaf and from the main columns

Fig. 5.37. A 3 mm wide by 40 mm deep groove is cut into the concrete using a hand held saw. Care has to be exercised to ensure a straight cut. The reduction in cross-sectional area and second moment of area ensures that cracking occurs here. Note that the main mesh is stopped here and a shorter piece having its main bars running along the line of the cut contributes to the weakening at such joints

Design charts

Design Chart 1
Plain C30 concrete

Flexural strength = 2·0 N/mm^2

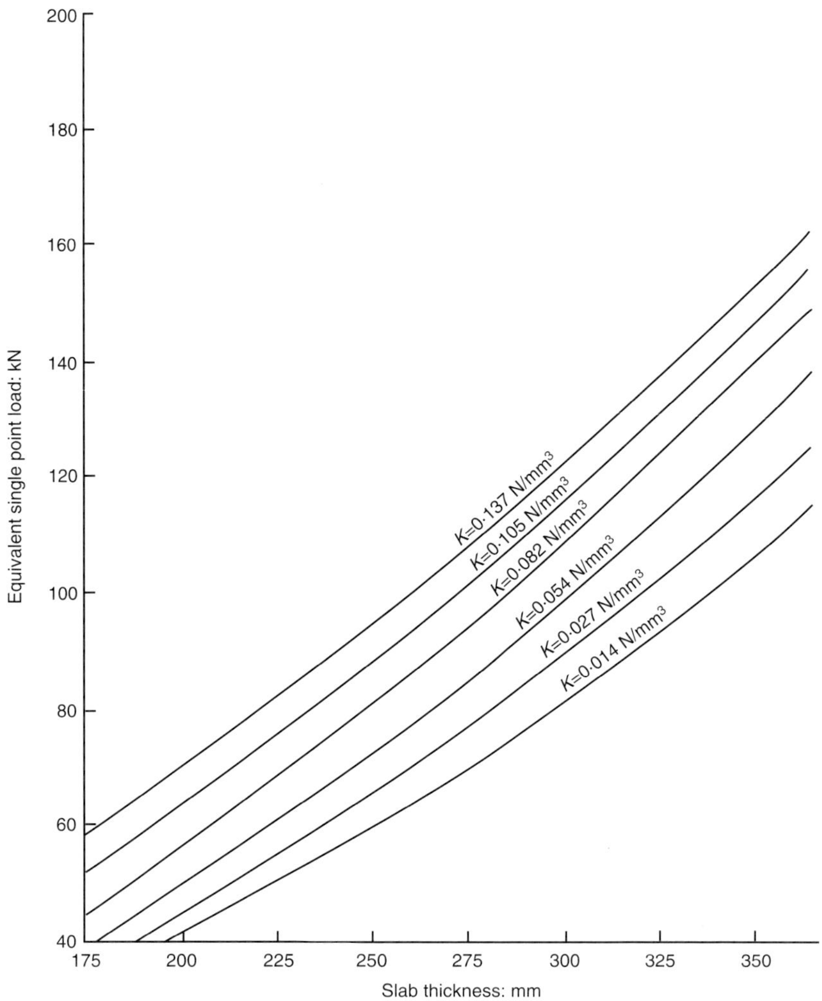

Design Chart 2
Micro-silica C30 concrete

Flexural strength = 2·4 N/mm^2

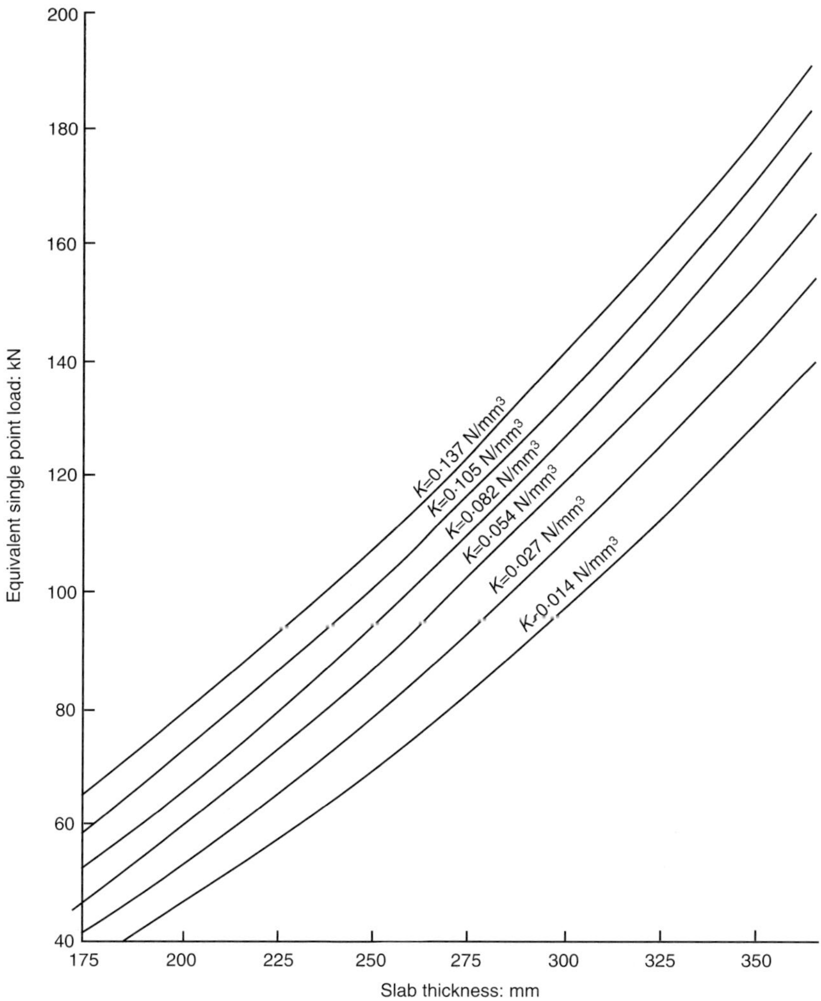

Design Chart 3
20 kg/m^3 ZC 60/1·00 steel fibre
C30 concrete

Flexural strength = 2·8 N/mm^2

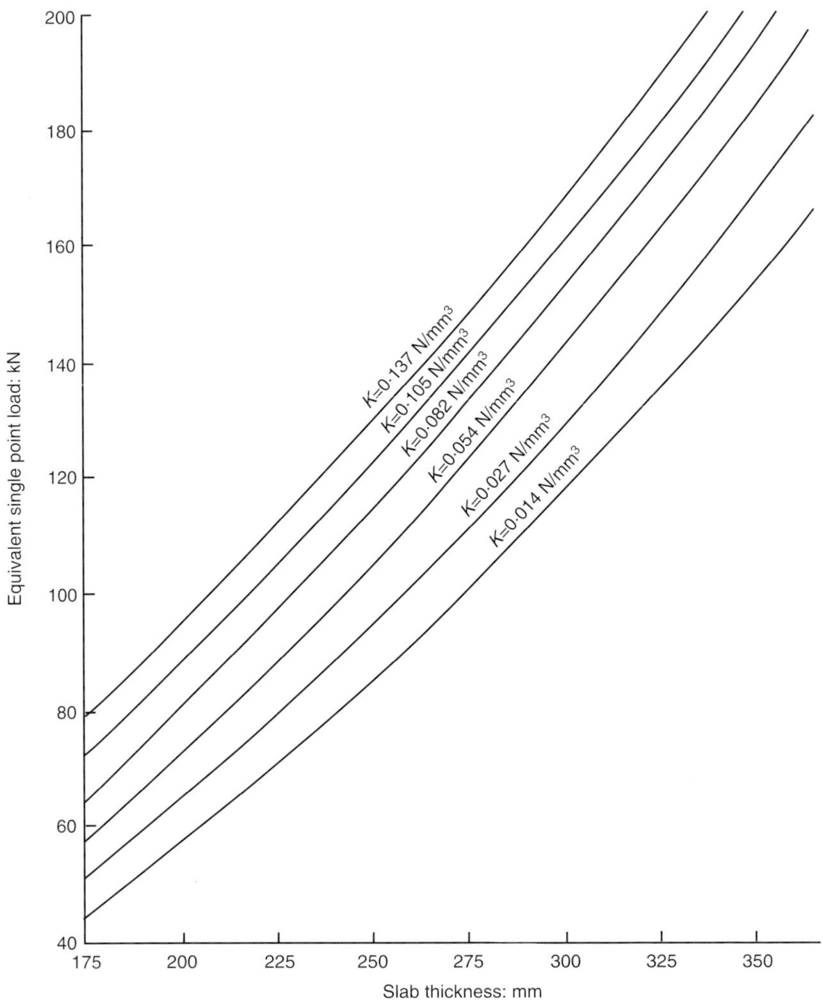

Design Chart 4
30 kg/m³ ZC 60/1·00 steel fibre
C30 concrete

Flexural strength = 3·2 N/mm²

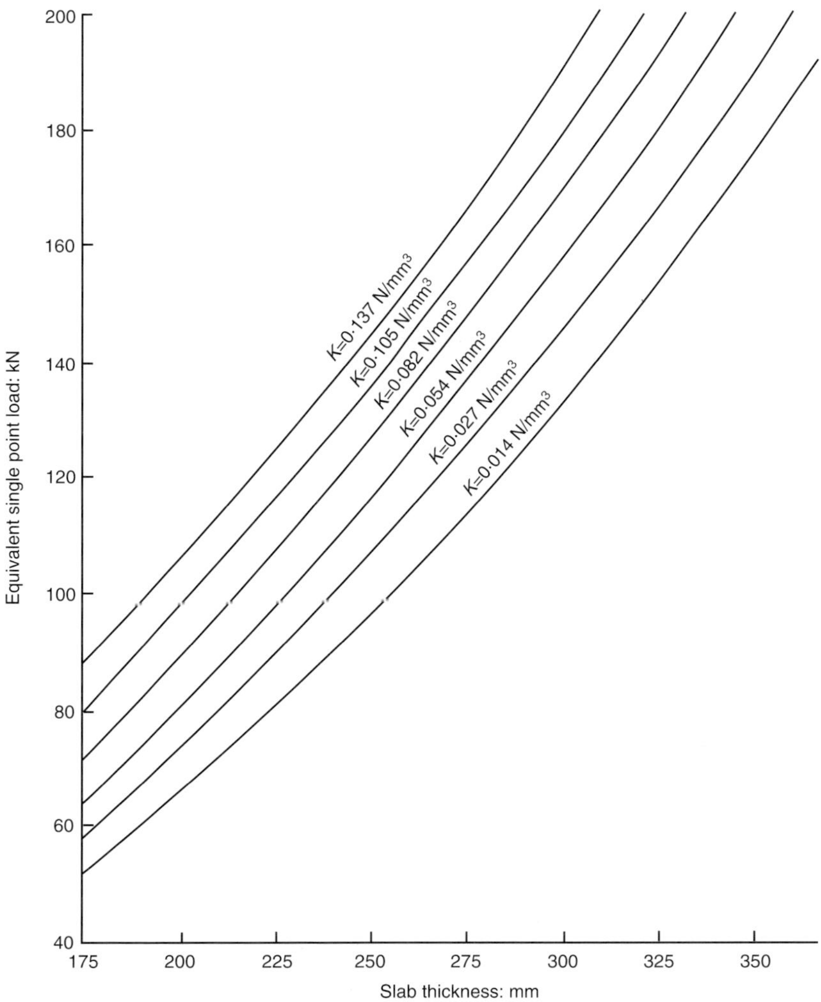

Design Chart 5
40 kg/m^3 ZC 60/1·00 steel fibre
C30 concrete

Flexural strength = 3·8 N/mm^2

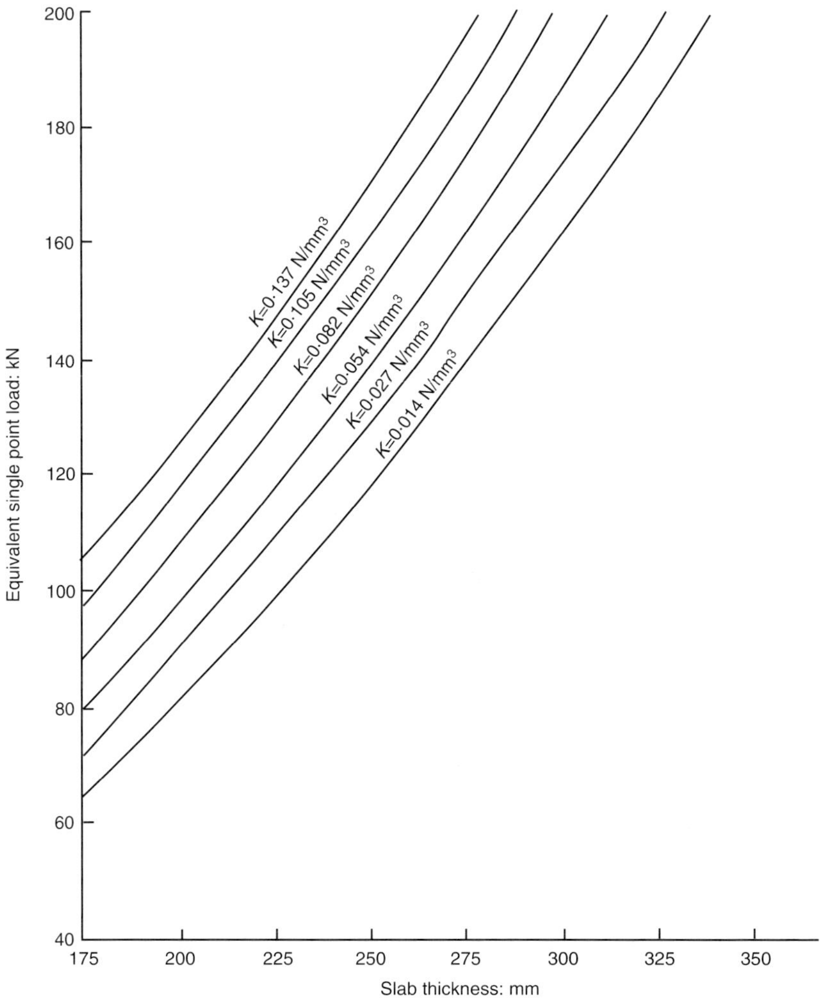

Design Chart 6
Plain C40 concrete

Flexural strength = 2·4 N/mm^2

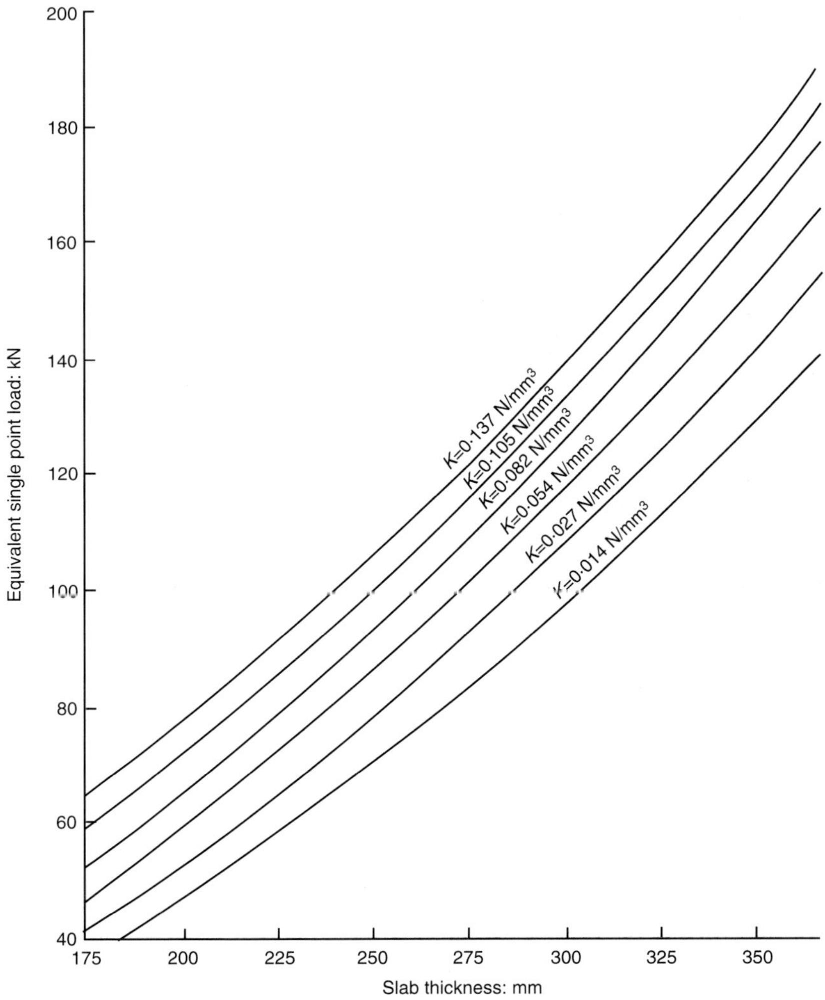

Design Chart 7
Micro-silica C40 concrete

Flexural strength = 2·8 N/mm²

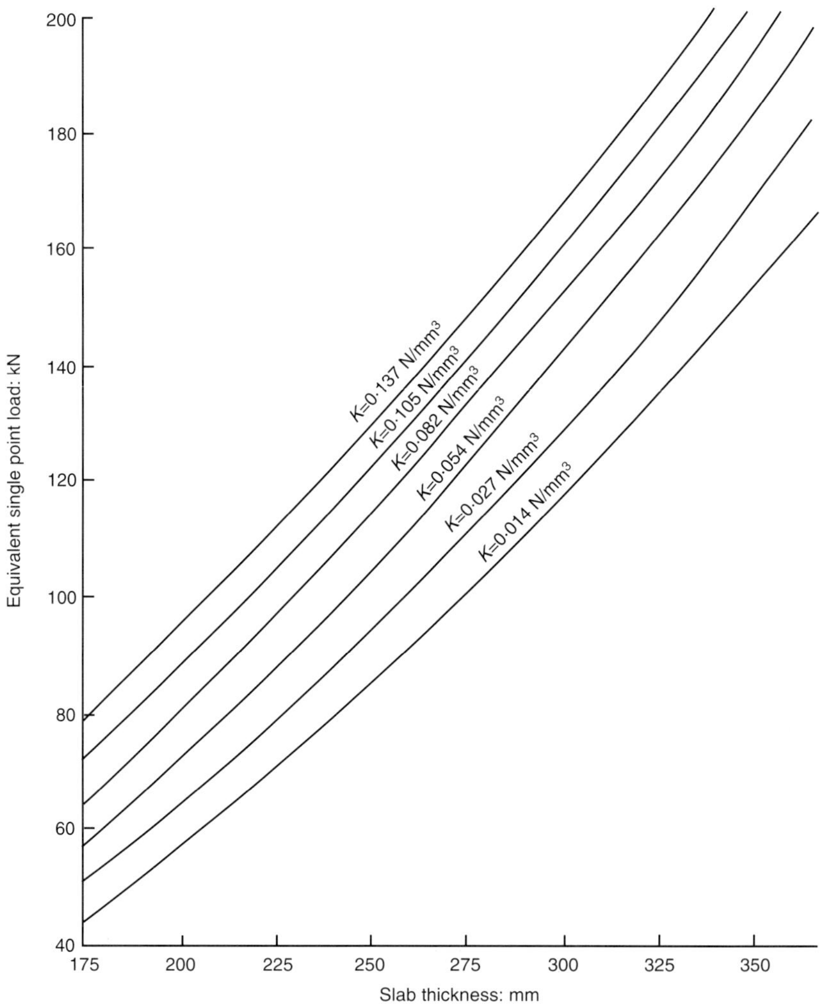

Design Chart 8
20 kg/m³ ZC 60/1·00 steel fibre
C40 concrete

Flexural strength = 3·2 N/mm²

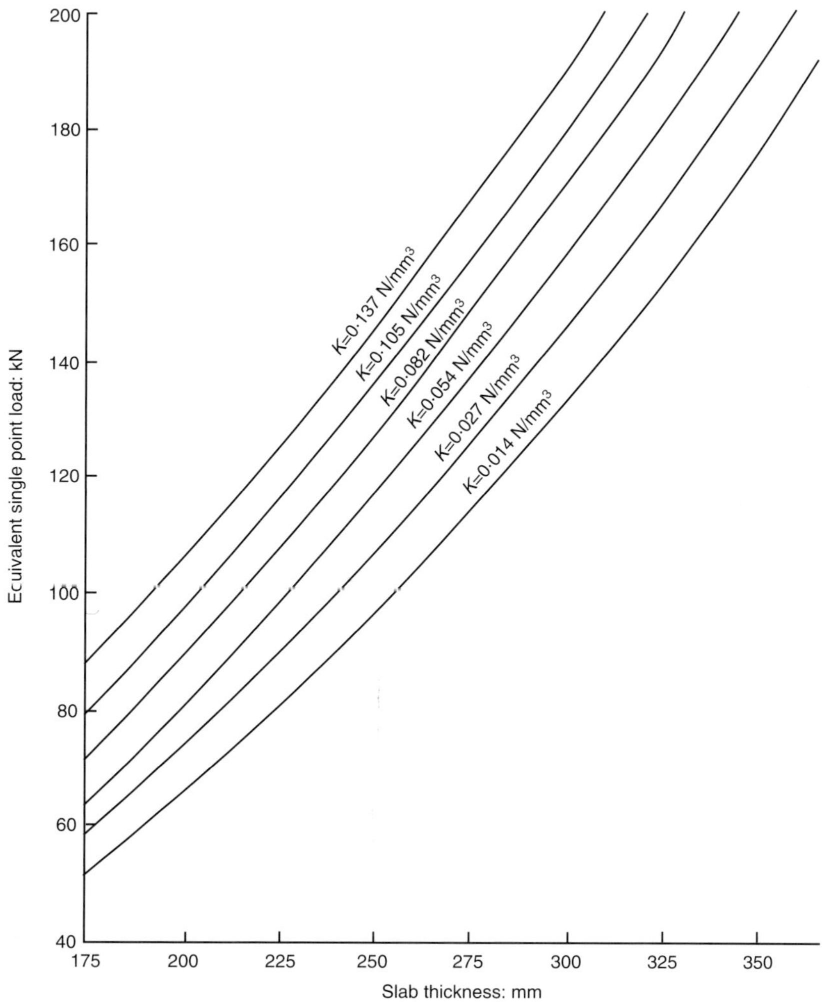

Design Chart 9
30 kg/m^3 ZC 60/1·00 steel fibre
C40 concrete

Flexural strength = 3·6 N/mm^2

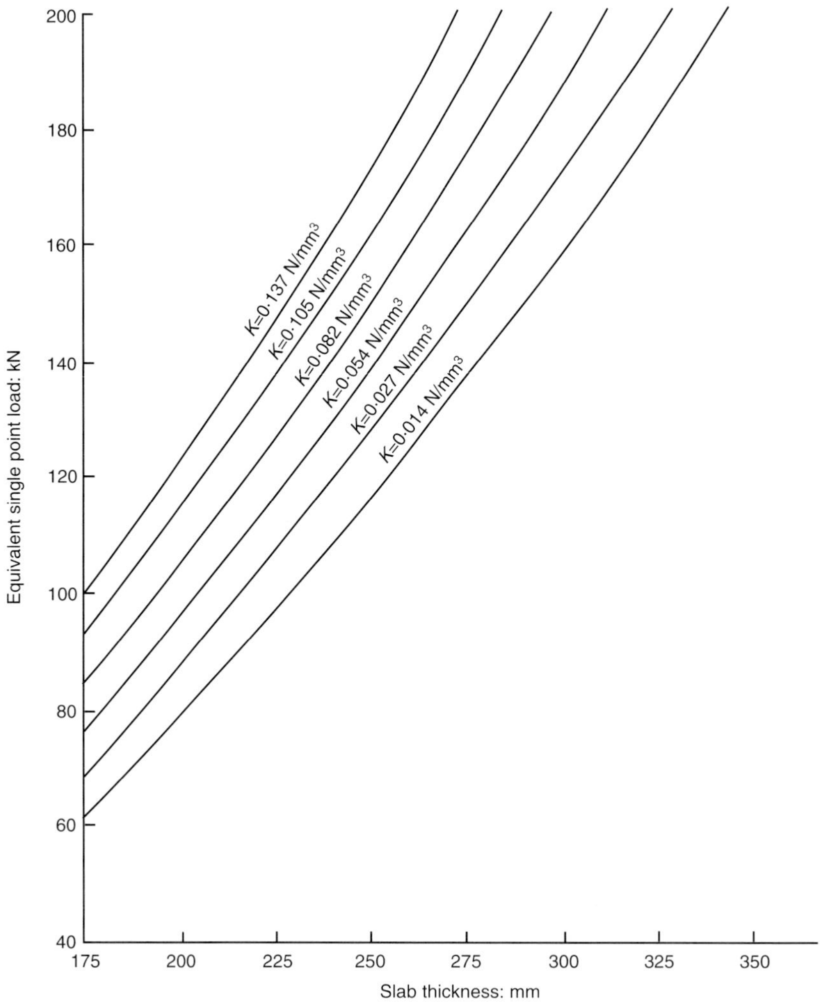

Design Chart 10
40 kg/m^3 ZC 60/1·00 steel fibre
C40 concrete

Flexural strength = 4·2 N/mm^2

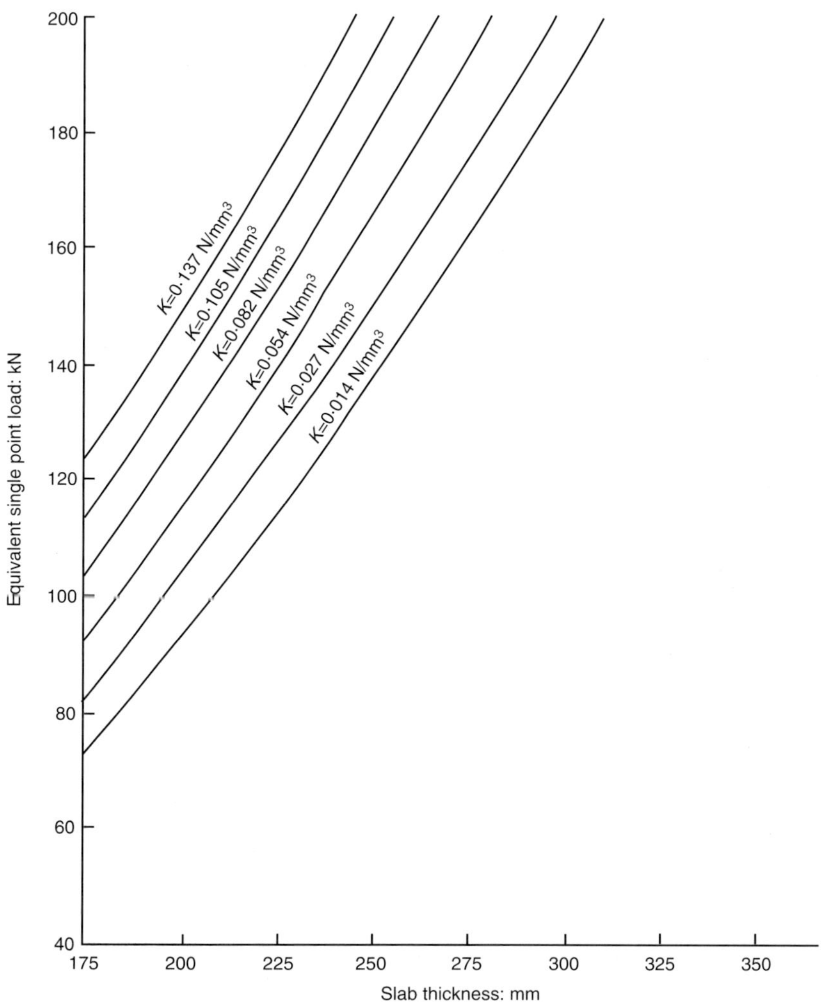

References

Chapter 1 Materials

1.1 British Standards Institution (1981). *Methods of Specifying Concrete*. BSI, London, BS 5328.

1.2 Deacon, R.C. (1988). *Concrete Industrial Ground Floors*. The Concrete Society/ British Industrial Truck Association/Storage Equipment Manufacturers' Association, Concrete Society Technical Report 34.

1.3 British Standards Institution (1992). *Aggregates from Natural Sources for Concrete*. BSI, London, BS 882.

1.4 Wright, P.J.F. (1965). *The Flexural Strength of Plain Concrete*. Road Research Laboratory, TRL, Crowthorne, Berkshire, UK, Technical Report TR67.

1.5 British Standards Institution (1982). *Concrete Admixtures*. BSI, London, BS 5075: Part 1.

1.6 British Standards Institution (1985). *Superplasticizing Admixtures*. BSI, London, BS 5075: Part 3.

1.7 Chandler, J.W.E. and Neal, F.R. (1988). *The Design of Ground Supported Concrete Industrial Floor Slabs*. BCA Interim Technical Note 11. British Cement Association, Crowthorne, Berkshire, UK.

1.8 British Standards Institution (1985). *The Structural Use of Concrete: Part 1: Code of Practice for Design and Construction*. BSI, London, BS 8110: Part 1.

1.9 British Standards Institution (1990). *Methods of Test for Soils for Civil Engineering Purposes*. BSI, London, BS 1377: 1990.

1.10 Bekaert (1990). *Industrial Floors with Dramix Steel Wire Fibre Reinforced Concrete*, NV Bekaert S. A. Zwevegem, Belgium.

1.11 Department of Transport. *Specification for Highway Works* Vol. 1, Road Pavements – Unbound Materials: Series 800. HMSO, London.

1.12 Department of Transport. *Specification for Highway Works: Road Pavements – Concrete and Cement Bound Materials*: Series 1000. HMSO, London.

1.13 British Standards Institution (1997). *Methods for Specifying Concrete Mixes*. BSI, London, BS 5328: Part 2: 1997.

1.14 Fibermesh Company (1985–1988). *Fibermesh Engineering Data Reports* Nos 1 to 7. Fibermesh Co., 4019 Industry Drive, Chattanooga, TN, USA.

1.15 British Standards Institution (1987). *In-situ Floorings: Part 2. Code of Practice for Concrete Wearing Surfaces*. BSI, London, BS 8204: Part 2 Sections 2.1 and 2.2.

1.16 Tatnall, P.C. and Kuitenbrouer, L. (1992). *Steel Fiber Reinforced Concrete in Industrial Floors*. NV Bekaert S. A., Zwevegem, Belgium.

1.17 (1987). Research Committee CUR (The Netherlands Centre for Civil Engineering Research and Codes). Directive No 10, *Design, Calculation and Placing of Industrial Floors with Steel Fibre Reinforced Concrete*.

1.18 TNO (1988). TNO Institute for Building Materials and Structures, Delft, Reports B88/607 and B88/751.

1.19 (1984). *Methods of Tests for Flexural Toughness of Steel Fiber Reinforced Concrete*. Japanese Society of Civil Engineers. JSCE – SF4.

1.20 British Standards Institution (1985). *Welded Steel Wire Fabric for the Reinforcement of Concrete*. BSI, London, BS 4483.

1.21 British Standards Institution (1978). *Cold Worked Steel Bars for the Reinforcement of Concrete*. BSI, London, BS 4461.

1.22 Deacon, R.C. (1986). *Concrete Ground Floors: their Design, Construction and Finish*. Third Edition. Cement and Concrete Association, Wexham Springs, Slough, UK.

1.23 Kettle, R.J. (1989). *Recent Research Developments in Abrasion Resistance*. Aston University, Birmingham, UK.

Chapter 2 Construction

2.1 Deacon, R.C. (1988). *Concrete Industrial Ground Floors*. The Concrete Society/ British Industrial Truck Association/Storage Equipment Manufacturers Association, Concrete Society Technical Report 34.

2.2 Snowdon Flooring (1994). *Technical Brochure*. Snowdon Flooring, North East Office, Houlst Estate, Walker Road, Newcastle upon Tyne, UK.

2.3 PCF Flow Line Slab Construction, John Kelly (Lasers) Ltd. and Fibrecon UK Ltd. (1992). Amorex, State-of-the-art Floor Construction. *World of Concrete Europe 92, 6–9 October 1992*, QEII Centre, London. British Precast Concrete Federation.

2.4 Soff-Cut International *Technical Brochure*. Soff-Cut International, Laymarston Distribution, The Rosary, Broad Marston Pebworth, Stratford Upon Avon CV37 8XY, UK.

Chapter 3 Loading

3.1 Deacon, R.C. (1988). *Concrete Industrial Ground Floors*. The Concrete Society/ British Industrial Truck Association/Storage Equipment Manufacturers Association, Concrete Society Technical Report 34.

Chapter 4 Design

4.1. Westergaard, H.M. (1947). New formulae for stress in concrete pavements of airfields. *Proceedings of the American Society of Civil Engineers* **73**, No 5, 127–138.

Chapter 5 Case Studies

5.1 Meyers, B.L. and Thomas, E.W. (1983). Elasticity, shrinkage, creep and thermal movement of concrete. In *Handbook of Structural Concrete* (eds. F.K. Kong, R.H. Evans, E. Cohen and F. Roll) Pitman, London, pp. 11-1–11-33.

Bibliography

American Concrete Institute (1982). *Concrete Craftsman Series. Slabs on Grade*. ACI, Detroit, p. 76.

Barnbrook, G. (1975). *Concrete Ground Floor Construction for the Man on Site*. Part 1: For the site supervisor and manager. Cement and Concrete Association, Slough, UK. Publication 48.035.

Barnbrook, G. (1976). *Concrete Ground Floor Construction for the Man on Site*. Part 2: For the floor layer. Cement and Concrete Association, Slough, UK. Publication 48.036.

Barnbrook, G. (1987). *Ground-Supported Concrete Floor Slabs*. Cement and Concrete Association, Slough, UK, Reprint May 1987.

British Standards Institution (1984). *Design loading for buildings. Part 1: Code of practice for dead and imposed loads*. BSI, London, BS 6399: Part 1, p. 11.

British Standards Institution (1985). *Structural use of concrete. Part 1: Code of practice for design and construction*. BSI, London, BS 8110: Part 1.

British Standards Institution (1985). *Specification for steel fabric for the reinforcement of concrete*. BSI, London, BS 4483, p. 8.

Chandler, J.W.E. (1982). *Design of Floors on Ground*. Cement and Concrete Association, Slough, UK, TRA/550, p. 22.

Chandler, J.W.E. and Neal, F.R. (1988). *The Design of Ground-Supported Concrete Industrial Floor Slabs*. British Cement Association, Slough, UK, BCA ITN11, p. 16.

Chaplin, R.G. (1972). *The Abrasion Resistance of Concrete Floors*. Cement and Concrete Association, Slough, UK, Publication 42.471, p. 9.

Chaplin, R.G. (1974). *The Regularity of Concrete Floor Surfaces: A Survey of Current Knowledge*. Construction Industry Research and Information Association, London, CIRIA Report 48, p. 31.

Chaplin, R.G. (1986). *The Influence of Cement Replacement Materials, Fine Aggregate and Curing on the Abrasion Resistance of Industrial Concrete Floor Slabs*. Cement and Concrete Association, Slough, UK, p. 32.

Chou, Y.T. (1983). Subgrade contact pressure under rigid pavements. *Proceedings of the American Society of Civil Engineers Journal of Transportation Engineering* **109**, No. 3, pp. 363–379.

Deacon, R.C. (1976). *High Strength Toppings for Floors, Including Granolithic*. Cement and Concrete Association, Slough, UK, Publication 48.010. Reprinted with amendments, 1984. pp. 7.

Deacon, R.C. (1986). *Concrete Ground Floors: Their Design, Construction and Finish*. Cement and Concrete Association, Slough, UK, Publication 48.034, p. 23.

Deacon, R.C. (1990). Fibres for floors. *Concrete* **24**, No. 4.

Face, A. (1982). A time bomb in floor tolerances? *Concrete Construction* **27**, No. 11, 865–869.

Face, A. (1984). Specification and control of concrete floor flatness. *Concrete International Design and Construction* **6**, No. 2, 56–63.

Hetenyi, M. (1946). *Beams on Elastic Foundations*. The University of Michigan Press. p. 255.

Hodgkinson, J.R. (1982). *Steel Reinforcement for Concrete Road Pavements*. Cement and Concrete Association of Australia, Sydney, Publication TN49, p. 16.

Hodgkinson, J.R. (1982). *Thickness Design for Concrete Road Pavements*. Cement and Concrete Association of Australia, Sydney, Publication TN46, p. 16.

Hubler, R.L. (1983). Industrial floors go superflat. *Concrete International Design and Construction* **5**, No. 12, 61–63.

JSCE (1987). *Recommendations for Design and Construction of Steel Fibre Reinforced Concrete*. Concrete library of JSCE, No. 3, June 1987.

Knapton, J. (1994). *The Structural Design of Heavy Duty Pavements for Ports and Other Industries*. British Precast Concrete Federation/British Ports Association, Leicester/London, p. 110.

Knapton, J. and Ellerton, J. (1994). *Single Pour Industrial Floor Slabs*. Newcastle University Ventures Ltd, Newcastle upon Tyne, UK.

Neal, F.R. and Judge, C.J. (1987). *Classes of Imposed Floor Loads for Warehouses*. Building Research Establishment Information Paper IP 19/87, Garston, UK, p. 4.

New Zealand Portland Cement Association (1980). *Design of Concrete Ground Floors for Commercial and Industrial Use*. NZPCA, Wellington, NZPCA IB0 26. p. 24.

Packard, R.G. (1976). *Slab Thickness Design for Industrial Concrete Floors on Grade*. Portland Cement Association, Skokie, PCA IS 195.01D, p. 16.

Panak, J.J., McCullough, B. and Treybig, J. (1973). *Design Procedure for Industrial Slabs*

Reinforced with Welded Wire Fabric. Wire Reinforcement Institute, Austin, USA, p. 45.

Panak, J.J. and Rauhut, J.B. (1975). Behaviour and design of industrial slabs on grade. *Journal of the American Concrete Institute* **72**, No. 5, 219–224.

Pateman, J.D. (1984). Specifying and constructing flat floors. *Concrete* **18**, No. 3, 7–9.

Rice, P.F. (1957). Design of concrete floors on ground for warehouse loadings. *Journal of the American Concrete Institute* **54**, No. 2, 105–113.

Ringo, B.C. (1978). Design, construction and performance of slabs-on-grade for industry. *Journal of the American Concrete Institute* **75**, No. 11.

Ringo, B.C. and Steenkers, R. (1982). *Industrial Floor Slabs: A Thickness Solution. ACI Computer Program Series / COM-1 (83). Design of Structural Concrete*. American Concrete Institute, Detroit, pp. 1–11.

Spears, R.E. (1983). *Concrete Floors on Ground*. Portland Cement Association, Skokie, Publication EB075.02D, p. 28.

Teller, L.W. and Sutherland, E.C. (1943). The structural design of concrete pavements. *Public Roads* **23**, No. 8, 167–212.

Timms, A.G. (1964). Evaluating subgrade friction reducing mediums for rigid pavements. *Highways Research Board*, Washington D.C., No. 60, p. 48–59.

Westergaard, H.M. (1925). Computation of stresses in concrete roads. *Proceedings of the 5th Annual Meeting of the Highways Research Board*, vol. 5, Part 1, p. 90–112. Highways Research Board, Washington D.C.

Westergaard, H.M. (1926). Stresses in concrete pavements computed by theoretical analysis. *Public Roads* **7**, No. 2, pp. 48–56.

Westergaard, H.M. (1933). Analytical tools for judging results of structural tests of concrete pavements. *Public Roads* **14**, No. 10, 185–188.

Westergaard, H.M. (1943). Stress concentrations in plates loaded over small areas. *Transactions of the American Society of Civil Engineers* Paper No. 2197.

Westergaard, H.M. (1947). New formulae for stress in concrete pavements of airfields. *Proceedings of the American Societry of Civil Engineers* **73**, No. 5, 687–701.

Ytterberg, R.F. (1987). Shrinkage and curling of slabs on grade. Part 1: drying and shrinkage. *Concrete International Design and Construction* **9**, No. 4, 21–31.

Ytterberg, R.F. (1987). Shrinkage and curling of slabs on grade. Part 2: warping and curling. *Concrete International Design and Construction* **9**, No. 5, 54–61.

Ytterberg, R.F. (1987). Shrinkage and curling of slabs on grade. Part 3: additional suggestions. *Concrete International Design and Construction* **9**, No. 6, 72–81.